1

IB GEOGRAPHY
Internal Assessment

IB GEOGRAPHY
Internal Assessment

The Definitive Geography [HL/SL] IA
Guide For the International Baccalaureate [IB] Diploma

Joanna Piotrowska
Alexander Zouev

Zouev IB Diploma Publishing

Published 2023

Printed by Zouev IB Diploma Publishing

ISBN 978-1-9163451-8-8, paperback.

TABLE OF CONTENTS

PART I
THE IB GEOGRAPHY IA GUIDE

1. GENERAL INTRODUCTION

The geography Internal Assessment consists of fieldwork written report. It has different weightings for SL and HL students – **25% and 20%** accordingly, but the criteria and maximum score of 25 points are the same for each level. The fieldwork should be conducted on a **local scale**, which means that the investigated area should be relatively small and situated in the surroundings of your school. Topic of investigation must be **linked to the syllabus** – you need to base your IA on the knowledge gained during geography course. Report should be based on primary information, which is quantitative or qualitative data collected by students and supplemented by secondary information, which involves data gathered from already existing resources[1].

Data for geography IA can be collected in the group. It is relatively nice idea as there is simply a lot to collect and write about. However, it will require good communication and cooperation between group members as methodology and data analysis are crucial in the fieldwork report. Below there are some tips on IA group collaboration:

a) **Plan** – think about what data you need for your IA and how much time will the collection of each type require. Divide tasks equally and set specified deadlines. If more people are responsible for gathering one type of data make sure to agree on the same method (for example, if you count floors of buildings, make it clear if you treat ground floor as 0 or 1).

b) **Communicate** – if you realize you won't be able to provide your data on time, inform other members of the group and tell them when they will be available – it's better than saying nothing. However, you should obviously avoid missing deadlines in group assignments.

c) **Describe** – make sure other members of the group can understand your data. Write a description of how you collected them, include information about which websites and maps you used – it will be important for methodology chapter. Provide data in tables that are clearly described.

You can only cooperate in the group for data collection and exchange of possible presentation methods. **The writing of the report must be done on your own.**

---

[1] All secondary information must be referenced using system of your choice to avoid plagiarism.

There is a **2,500 word limit** for geography IA, which doesn't include: title page, acknowledgments, contents page, titles and subtitles, references, footnotes (up to 15 words each), map legends and/or keys, labels (up to 10 words each), tables (of statistical or numerical data, categories, classes or group names), calculations, appendices (containing only raw data and/or calculations, which should be representative). Although the limit might seem a lot, it is not that hard to exceed it and as the teacher is not obliged to read anything after 2,500[th] word, you need to keep your IA concise with maximum content in minimum words. The nice method of keeping your eye on wordcount is to copy all text that should be counted into separate word document, while leaving all the other parts in the proper IA file.

Internal assessment means that yours and your fellow students' reports will be graded by your teacher and some of them might be chosen to go through moderation by IB examiners. Hence, make sure to bear in mind your teacher's suggestions after initial proposal and first draft and consult your concerns with them, as they will have major influence on your final score.

2. GRADING CRITERIA

The IA is marked against 6 criteria to give a total of 25 marks and there is a suggested wordcount for each section[2]:

CRITERION	NAME	MAX MARK	SUGGESTED WORDCOUNT
A	Fieldwork question and geographic context	3 marks	300 words
B	Method(s) of investigation	3 marks	300 words
C	Quality and treatment of information collected	6 marks	500 words
D	Written analysis	8 marks	850 words
E	Conclusion	2 marks	200 words
F	Evaluation	3 marks	300rds

2.1 Fieldwork question and geographic context

- Fieldwork question needs to be precise – name the area and subject of your investigation (e.g. To what extent does the city centre of [**city**], [**country**], fit the **central business district (CBD) model**, based on **[number] characteristic features** of CBD?). You need to be able to answer the question using primary information (data collected by you and/or your group in the fieldwork). Your work should only contain information relevant in context of your question, so it's vital to formulate it correctly.

- You might include **hypothesis** based on some secondary information or general observations. Using example above, you could name the characteristic features mentioned in the fieldwork question and state if you expect to observe them in the chosen city.

---

[2] You don't have to stick to suggested wordcounts for sections, however they are useful tool to obtain right proportions in your IA.

- For the **geographic context** you can briefly describe the area you investigate, including some relevant historical, economic, cultural or landscape facts and reason why you chose particular region. You should include general map of a country with highlighted and named region of investigation (e.g. particular city or river).

- **Link to the syllabus** should be well described – name particular chapters/sub-chapters and information contained within them to which your report refers to.

2.2 Method(s) of investigation

- This section refers to what you did during the fieldwork, so it's best to think about it before you start the investigation. You will need to describe how you collected the data and why you chose particular method. You should justify your actions to show that you understand methods of the investigation.

- For each type of data you should first consider if it is **qualitative or quantitative**. To collect qualitative data you observe something and for quantitative data you count or measure something.

- Then you should think of the most appropriate **sampling method**. There are a few to choose from:

  a) **random sampling** – applies when you randomly choose objects for your investigation and every object has an equal chance to be selected.

  b) **systematic sampling** – applies when you choose objects according to a certain pattern, so that objects under investigation are evenly distributed or you make a measurement in regular time or spatial interval.

  c) **stratified sampling** – applies when you choose objects according to a certain proportion, for example if you survey a population that consists of 40% males

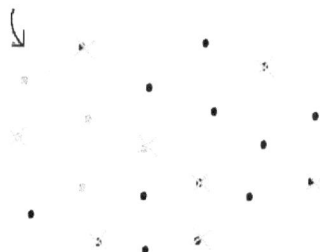

15

and 60% females, you should have 40% of answers from males and 60% from females who are chosen using random sampling.

d) **quota sampling** – it's basically the same as stratified sampling, only the sampling method used for choosing males and females from the example above is not random, but for example systematic.

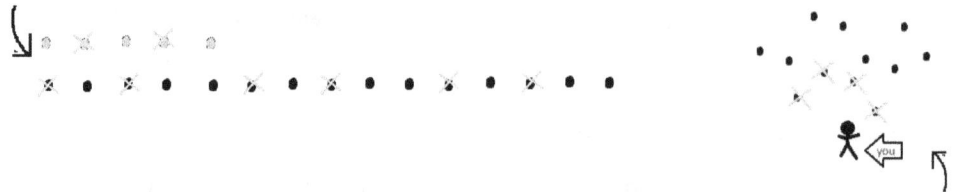

e) **convenience sampling** – applies when you choose objects from your closest environment, so your chosen group might be very specific and biased.

- If you find it appropriate you could prepare a map showing the objects chosen in spatial context, which would confirm which sampling method you used.
- You should think of **other relevant aspects** that could have impact on your results, for example if the streets chosen for investigation were of the same length, if the apartments chosen to compare land values were furnished/unfurnished or ready to live in/for renovation and if ground floor counts as 1 or 0.
- If you processed your data using some **formulas**, methodology chapter is a good moment to describe it as well.
- If you collaborated as a group for data collection, you can briefly mention the division of work at the beginning of the chapter. After that all of the above information should be stated as undertaken steps.
- You can write in both first and third person, however especially if you collected data in a group, it might be a better option to use third person writing as you didn't in fact do all the steps yourself.

2.3 Quality and treatment of information collected

- This criterion is slightly connected to methodology as it emphasizes that collected data should be **relevant** in context of your fieldwork question and allow sufficient analysis by its **quantity and quality**.

- It also refers to presentation of your data, so to obtain high score it's good to show a variety of techniques, including different graphs, diagrams, maps, tables, statistical tests, annotated photographs or images. For example:

a) **Column diagrams** – you can use them to display results for different **categories**. The horizontal axis should be labelled with names of categories and the vertical axis should be labelled with values of interest. You can adjust these diagrams for your needs for example:

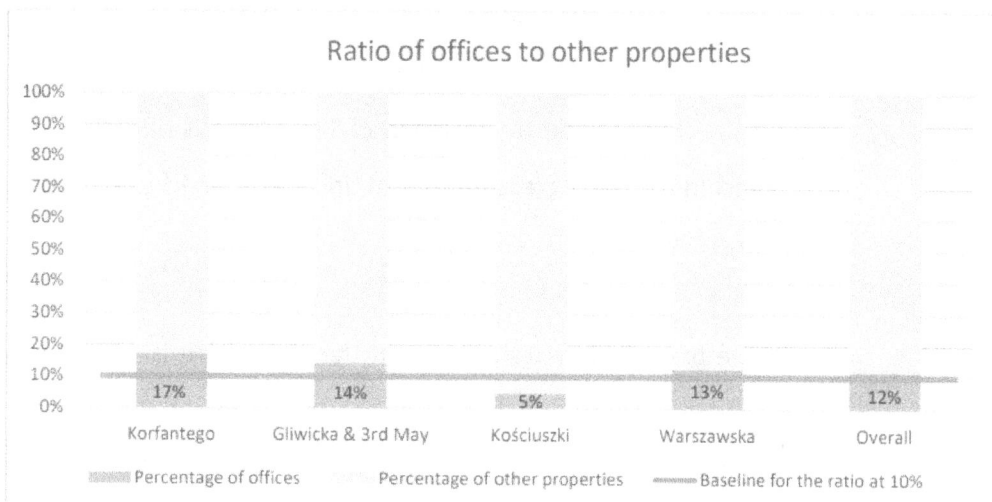

Ratio of offices to other properties

17%	14%	5%	13%	12%
Korfantego	Gliwicka & 3rd May	Kościuszki	Warszawska	Overall

▨ Percentage of offices   Percentage of other properties   ▬ Baseline for the ratio at 10%

- The graph below has street names as categories on horizontal axis but only percentage on vertical axis as it shows the ratio. **Baseline** is included to clearly show which streets meet the assumption that the ratio will be higher than 10%.

- The columns on the graph below represent buildings in increasing distance from the centre and the vertical axis shows their heights measured in floors. The **linear fit** was applied as it was expected that the height will be decreasing with distance from centre. However the R^2 value showed that data weakly fit the model[3] – this

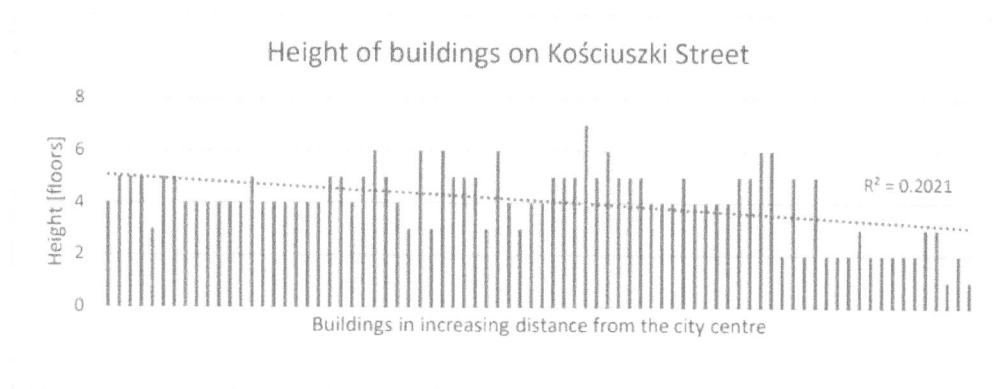

Height of buildings on Kościuszki Street

$R^2 = 0.2021$

Height [floors]

Buildings in increasing distance from the city centre

---

[3] R^2 and other statistics mentioned will be explained in next section of this guide.

information and explanation why this occurred supported by geographical knowledge and maybe secondary data is needed to score marks in the written analysis (next criterion discussed).

b) **Linear graphs** – both axes should be labelled with name and unit of the variable and values should be shown. **Horizontal axis** is for **independent variable** – the one that you choose to see how it influences the other one (if possible there should be even spaces between values, however this might be hard for fieldwork data) and **vertical axis** is for **dependent variable** – the one that changes due to influence of the previous one. You can almost always include some **regression model** and the **R** or **R²** values to comment on the correlation.

- On the graph below distance from the city centre is independent variable and price is dependent variable, as it was meant to show the influence of distance from the city centre on price of apartments for renovation. Linear fit and R^2 value are shown.

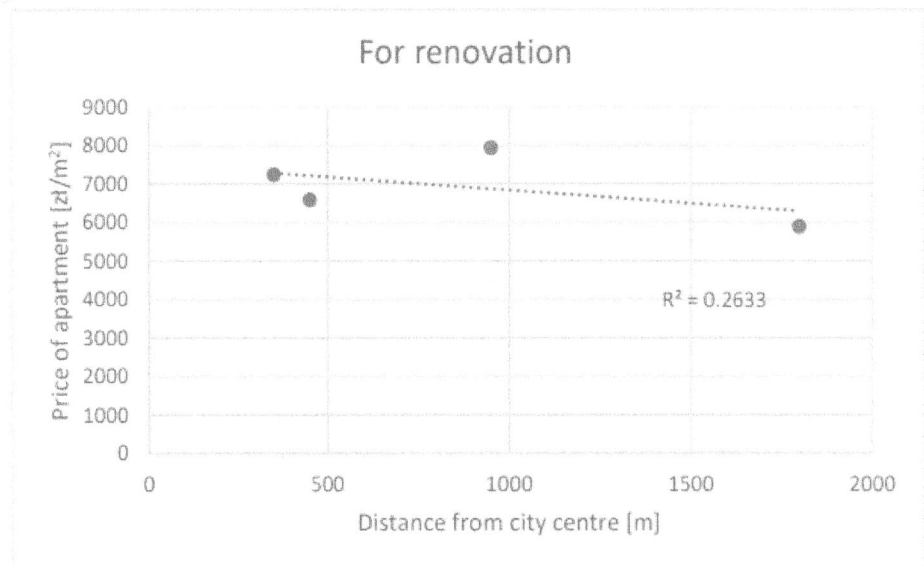

For renovation scatter graph, y-axis: Price of apartment [zł/m²] from 0 to 9000, x-axis: Distance from city centre [m] from 0 to 2000, $R^2 = 0.2633$

c) **Pie charts** – these can be used to show proportions.

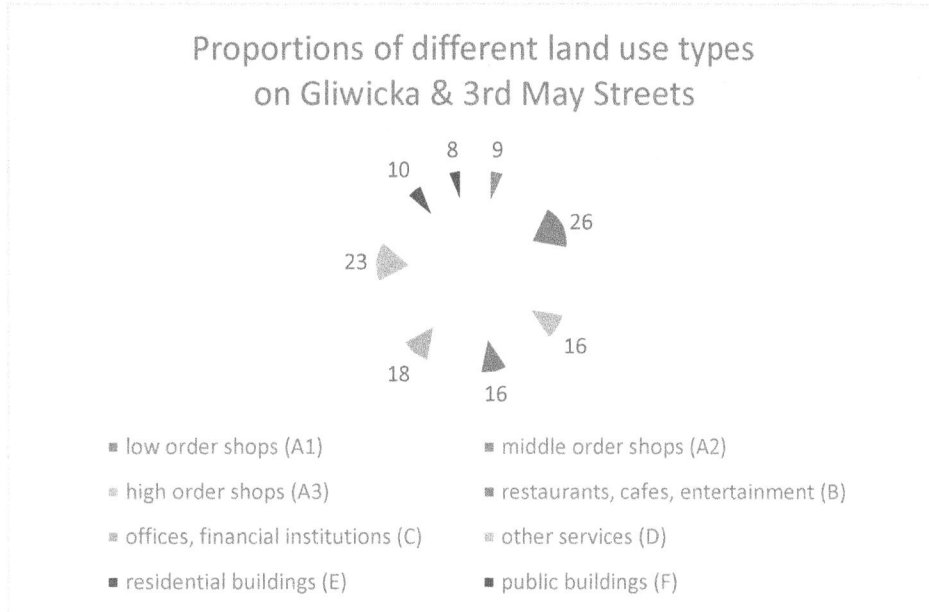

Proportions of different land use types
on Gliwicka & 3rd May Streets

- low order shops (A1)
- middle order shops (A2)
- high order shops (A3)
- restaurants, cafes, entertainment (B)
- offices, financial institutions (C)
- other services (D)
- residential buildings (E)
- public buildings (F)

d) **Maps** – all of them should include: **key/legend, north direction and scale**. You can use them to show location of investigated area (general map should be included in background information), but also distribution patterns.

- The map below is a general map showing location of investigated area (country and region with borders together with city where the fieldwork took place).

- The map below represents systematic sampling method (the streets of approximately equal lengths were chosen starting in the city centre and emerging in four world directions – north, east, south and west).

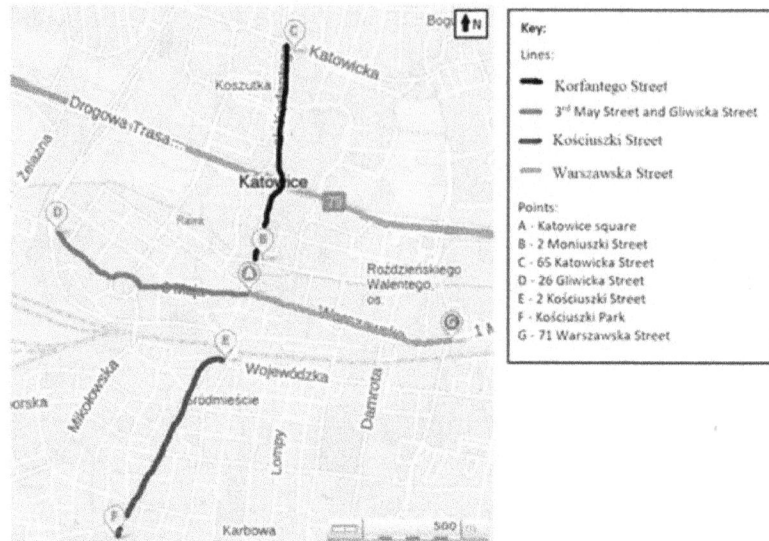

- The maps below show distributions. The first one represents distribution of shops within chosen area, while the second one shows distribution of all types of land uses on investigated streets. From this kinds of maps conclusions about clustering of certain types of land uses could be drawn.

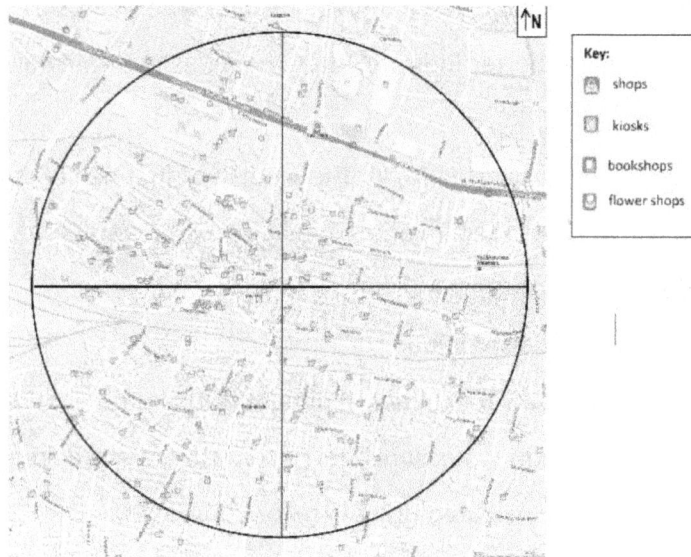

Key:
- shops
- kiosks
- bookshops
- flower shops

e) **Tables** – provide a good way to clearly show data used for producing diagrams. If you have a lot of raw data that aren't really needed in the main body, you can put tables in the appendix.

f) **Photographs** – might be a good way to visualise some trends or anomalies.

- You can use excel to prepare graphs and diagrams and simple graphic programmes for edition or preparation of maps. If you use any image as a base for your drawing, you should cite it in your work.

- All of the above should be labelled underneath (Figure 1,2,3...) except for tables which should be labelled above (Table 1,2,3...). Remember about word limit – 10 for each label, so that it's not included in the overall wordcount.

2.4 Written analysis

- Simply presenting collected data on the graphs is not enough to score high in geography IA. You need to show that you understand what you include in your work by analysing each figure.

- You should refer to any **patterns and trends** seen and try to explain them using geographical knowledge from the syllabus.

- If there is any **anomaly or outlier**, try to find a reason for those.

- When possible try to include **statistics**, such as:

a) **measures of central tendency:**

- mean – average value of your results;

21

- median (Q_2) – middle value of your results, when you sort them from lowest to highest, can be better indicator than mean, as extreme values don't influence it so much;

- quartiles – lower quartile (Q_1) is the median of first half of your results and higher quartile (Q_3) is the median of second half of your results;

- mode – value occurring most often in your data set.

b) measures of dispersion:

- range – maximum value minus minimum value in the data set;

- standard deviation – measures how much the data are spread around the mean value, it can be calculated using Excel or your GDC;

- interquartile range (IQR) – it is calculated from the formula $Q_3 - Q_1$ and gives the range within which 50% of data closest to median are located. It can be used to check for outliers – any value below $Q_1 - 1.5 \times IQR$ or above $Q_3 + 1.5 \times IQR$ can be considered outlier and doesn't have to be included in calculations.

c) trendline – can be displayed on your graph in Excel. Using this you can determine if the correlation between your variables is positive or negative, linear, exponential, polynomial or logarithmic. After adding the trendline you can proceed with statistics to R and/or R^2 values.

d) R (Pearson correlation coefficient) – can be calculated by your GDC. It takes values between -1 and 1 and shows you how strong the correlation between two variables is. In general, the more steep the line is, the stronger the correlation will be.

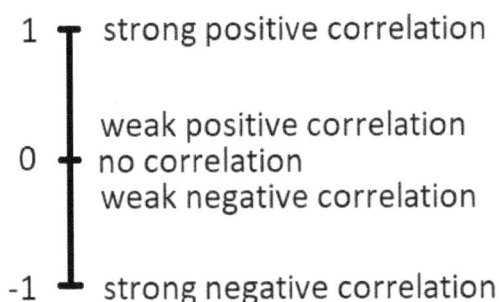

```
1 ┬ strong positive correlation
  │
  │ weak positive correlation
0 ┼ no correlation
  │ weak negative correlation
  │
-1 ┴ strong negative correlation
```

e) R^2 (coefficient of determination) – can be displayed by Excel on your graph. It takes values between 0 and 1 and is used to determine how well the data fit the statistical model applied.

1 ⊤ data fit the regressioon model perfectly

0 ⊥ data don't fit the regression model applied

straight line - even distribution

Lorenz curve - actual, not even distribution

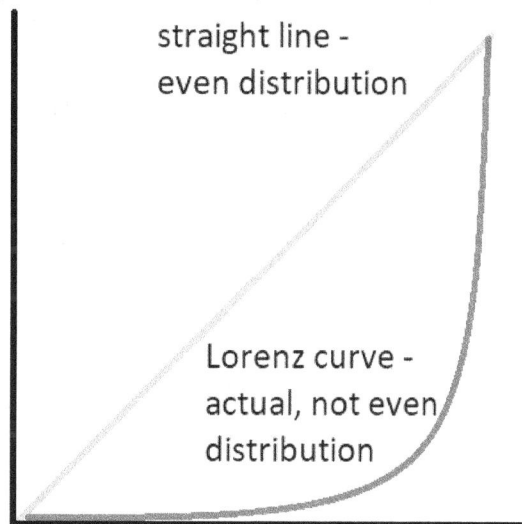

f) **NNI (nearest neighbour index)** – it is a formula that allows you to distinguish between clustered, random and regular distribution patterns.

$$NNI = 2\bar{D}\sqrt{\frac{N}{A}}$$

where:

$\bar{D}$ - average distance between each point and another point closest to it;

N – number of points under investigation;

A – size of area under investigation.

It takes values between 0 and 2.15 which mean the following:

g) **Gini coefficient and Lorenz curve** – can be used to show inequality, Gini coefficient takes values from 0 to 1, where 0 means equal distribution and 1 indicates total inequality. It is calculated using Lorenz curve, by dividing the area between straight line and Lorenz curve by total area under the straight line.

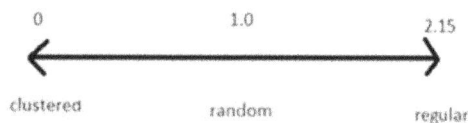

| 0 | 1.0 | 2.15 |
| clustered | random | regular |

h) Chi-square test – can be conducted when you have categorical data. It's aim is to show whether there is a significant difference between observed and expected situation. It can be used in some ecological tests to check if presence of one species is dependent upon presence of the other one or to check whether some data for countries would be equal or varied. The use of chi-square test slightly varies for specific examples, but in general you need to state null and alternative hypotheses, make tables of observed and expected frequencies and then apply the formula

$$x^2 = \Sigma \frac{(O - E)^2}{E}$$

where:

x^2 – chi-square statistic;

O – observed frequency;

E – expected frequency.

You need to calculate degree of freedom for your specific case and then find critical values – if your calculated x^2 value is higher than critical value for 0.05, it means that there is less than 5% probability that null hypothesis is correct.

- You conduct the analysis in order to be able to answer the fieldwork question, so all of it should directly relate to it and the hypotheses. When you include statistics, briefly describe what they are in general, comment on the values you obtain and mention **why they are important** in context of your investigation.

2.5 Conclusion

- In this section you should simply **summarize your results** in a concise way. Write which of your assumptions were correct and which weren't and refer to parts of data treatment or analysis that show it the best (include some numerical data as well).

2.6 Evaluation

- This criterion requires you to write about **weaknesses and errors** of the investigation. You need to think again about the methodology – perhaps in some cases different sampling method would make your data more accurate. Maybe during analysis you found out that smaller area of investigation would show different pattern than the one you chose. If the subject of your investigation could be affected by weather, think if all data were collected in the same conditions. You should also suggest **possible improvements**.

- Along with weaknesses your investigation should have some **strengths**. These can be very simple like not complicated method and easy access to required equipment, up-to-date data thanks to collecting them on your own instead of using internet or some of the sampling methods that were chosen correctly.

- Last part of evaluation consists of **possible extensions** – think about what more you could do. Are there any more features of the CBD that could be verified in your city? Are there other characteristics of the river flow that could be investigated by your group?

3. IA REPORT STRUCTURE

The information that you need to include to meet abovementioned criteria must be somehow structured. Your fieldwork report will look clearly if chapters and subchapters are numbered. There are also some formal requirements: you should use standard margins, well readable font like Times New Roman or Calibri of size 11-12 and line spacing of 1.5-2. You should also number pages in your IA. Below there is a sample structure of the geography IA.

Title page:

GEOGRAPHY FIELDWORK WRITTEN REPORT

Geography SL/HL

Fieldwork question

Session:

Personal code:

Wordcount:

Declaration: I confirm that this work is my own work and is the final version. I have acknowledged each use of the words or ideas of another person, whether written oral or visual.

Table of contents:

Main body:

1. **Introduction**

   1.1. **Geographical context**

   General information and map showing location of investigation.

   1.2. **Fieldwork question**

   Rewrite your fieldwork question from title page in the main body.

   1.3. **Hypothesis**

   Predicted answer to your question, you can also give more insight of what you are going to investigate.

   1.4. **Methodology**

   Step by step description of what was done as a fieldwork.

1.5. Link to the syllabus

Mention chapters/subchapters of the syllabus to which your IA refers to and describe the connections briefly.

2. Data analysis

Present your data in tables, diagrams and maps, include statistics and analyse them. To make it clear which data address which feature, it's best to add subchapters in this section.

2.1. Feature 1

2.2. Feature 2

2.3. Feature 3

...

3. Evaluation

3.1. Weaknesses and improvements

Weaknesses/limitations	Impact on the results	Possible improvements

3.2. Strengths

3.3. Possible extensions

4. Conclusion

5. Bibliography

6. Appendix

4. POSSIBLE TOPICS

Topic of your investigation is very much dependent on the environment you live in, as fieldwork is supposed to be at a local scale. Think about what is in the syllabus and try to find some common features in your surroundings. In general it's best when you can choose a subject that you expect to fit a certain model, which has features that you can measure yourself. The most commonly chosen subjects of investigations refer to:

- **Urban environments** - as this option gives many possibilities – you can investigate if your own or neighbouring city possess features of smart, sustainable or eco-city, if the city centre fits the CBD model or if the urban heat island effect occurs. You could also reflect on different kinds of pollution levels in the city or certain socio-economic variables.

- **Freshwaters** – as this option allows to measure fluvial characteristics in different parts of the river to check if they change with distance downstream according to the Bradshaw's model or to investigate the impact of different engineering schemes on flow characteristics. Influence of different factors on river/wetland biodiversity could also be a possibility for an IA topic.

Other interesting topics may emerge from the following ideas:

- **Oceans and coastal margins** – you could investigate the influence of human made structures on coastal erosion or social attitudes related to rising sea levels.

- **Tourism** – you could research its influence on environmental, social and economic aspects of a chosen area.

- **Ecology** – you could focus on influence of different factors on presence of a chosen species in a given area.

In general – choice of topic is only limited by your imagination, your teacher preferences, area where you live and the syllabus – a nice topic that can't be related to any part of a syllabus unfortunately can no longer be considered as a good topic.

5. FINAL TIPS

- You can write your IA in Word or Google docs. However, it's worth keeping somewhere a **copy** of it, as during writing you're going to paste many pictures and graphs and sometimes well-structured document can become completely chaotic. So to avoid unnecessary stress it's good to just have your document saved in version before you moved this picture 2 mm in wrong direction which magically ruined the whole structure.

- As a **base for your maps**, it might be better to use Targeo than Google maps – they have more schematic graphic when you zoom to buildings-street level and therefore it might be easier to use them.

- As an addition to Garrett Nagle's and Briony Cooke's Geography Coursebook (Oxford) there is **'Skills for IB geography'** booklet available, which includes some very specific statistics and data presentation methods, that might be useful for your IA.

- Try to **have fun** during data collection – geography IA gives you an opportunity to look at your surroundings from a different, more scientific perspective!

- Just to emphasize it one more time – everything you include in your work should be **relevant** in context of your research question. It is extremely important not to get sidetracked.

- **Good luck!**

PART II

SEVEN EXAMPLES OF EXCELLENT IB GEOGRAPHY IA

The IA featured in this section are all recently submitted IA that scored exceptionally well (band 7) after being moderated by the IBO. Where possible, actual examiner feedback has been included. The IA are presented in the exact same way as they were submitted, without any edits or changes to formatting. We do not retain the copyright of these IA, nor is this publication endorsed by the IBO. The Internal Assessments are being re-printed with the permission of the original authors.

1. EXAMPLE ONE (23/25)

Title: "How do the fluvial characteristics of Muang Ngam River, Chiang Mai, change with distance from the source?"

Author: Anonymous

Session: May 2022

Level: SL

Examiner's summary

Criterion A [2/3]:

The student has successfully stated a narrowly focused fieldwork question and justified the choice of location. They explored the research question by collecting primary and secondary data and presented a clear description of the geographical context of the river. The student also included a locational map with all the necessary components, and identified the link between the topic of the study and the areas of the geography syllabus. However, the explanations of the fieldwork's hypotheses were too general and brief to fulfill the criterion.

Criterion B [3/3]:

The student demonstrated a clear understanding of the methods selected for both primary and secondary data collection, including the tools required to apply them. They also justified the chosen methods and chose suitable techniques to collect quality data for subsequent analysis. Furthermore, the student included all necessary figures such as pictures, maps, or sample worksheets to explain the methods of the investigation. The date, time, and location where the data collection was conducted were also stated.

Criterion C [6/6]:

The student collected and presented enough relevant data to the fieldwork question, including annotated maps and graphs. All figures used by the student include a label, title, cardinal directions, and a scale. The sample size used by the student is sufficient for a detailed and in-depth analysis, specifically for a river. Maps used by the student are well-annotated and personalized to allow for quality treatment of collected data. The student used a variety of different techniques to present the collected data.

Criterion D [8/8]:

The student provided a thorough and detailed discussion about the collected data, correctly interpreted the trends and tendencies, and recognized and analyzed significant trends and patterns in the collected data. They also identified important outliers/anomalies and suggested their potential source. The descriptive techniques and applied statistical tests were appropriately selected for the data collected and the fieldwork question, and the student referred to geographical context and theory in their analysis. Their written analysis was relevant to the posed fieldwork question and assumed predictions/hypotheses.

Criterion E [2/2]:

The student's conclusion clearly answers the fieldwork question and is supported by the data collected and its analysis. They referred to their original predictions/hypotheses and compared them with the findings, and summarized the results of the fieldwork investigation. The final conclusion is well-supported by the collected data and its subsequent analysis.

Criterion F [2/3]:

The student effectively identified the weaknesses of the fieldwork methodology and correctly recognized the potential factors which could have affected the reliability of the collected data. However, the student did not write about any strengths of the methodology. The improvements suggested by the student are specific, doable, well-explained and justified.

Other requirements [0/0]:

The student has met the word limit requirement for the criterion. However, the pages are not numbered and there are no in-text citations for external sources of information, although a bibliography has been included.

"How do the fluvial characteristics of Muang Ngam River, Chiang Mai, change with distance from the source?"

Word Count: 2489 words

Table of contents

Abbreviations

(MNR) Muang Ngam River

(SRCC) Spearman's Rank Correlation Coefficient

Introduction

Research question

This study aims to answer the following fieldwork question:

"How do the fluvial characteristics of a Muang Ngam River change with distance from the source?"

Relation to syllabus

This investigation relates to Option A (Freshwater), specifically section 1 (Drainage Basin Hydrology and Geomorphology).

Location

The study is specific to the Muang Ngam River. The river flows through Mae Ai District in Chiang Mai Province, located in the North of Thailand (Figure 1.1).

Figure 1.1: Annotated map of Thailand and its location in Southeast Asia

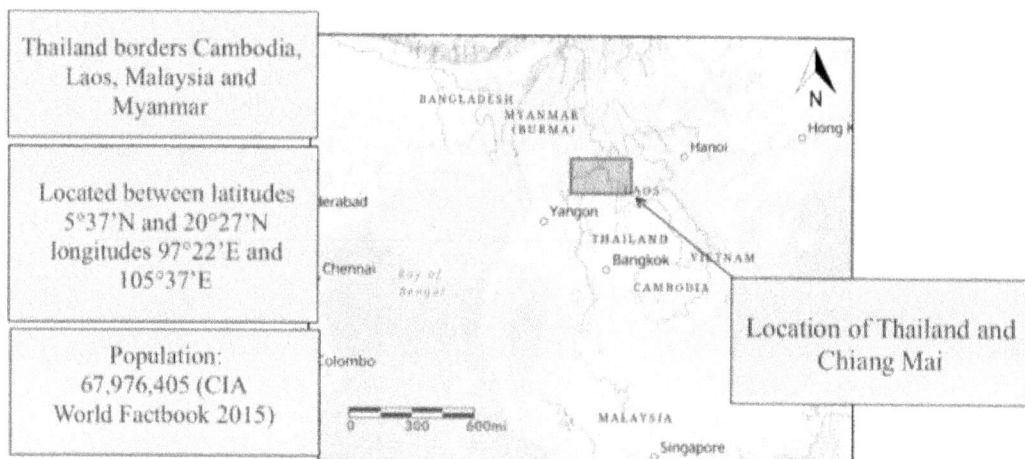

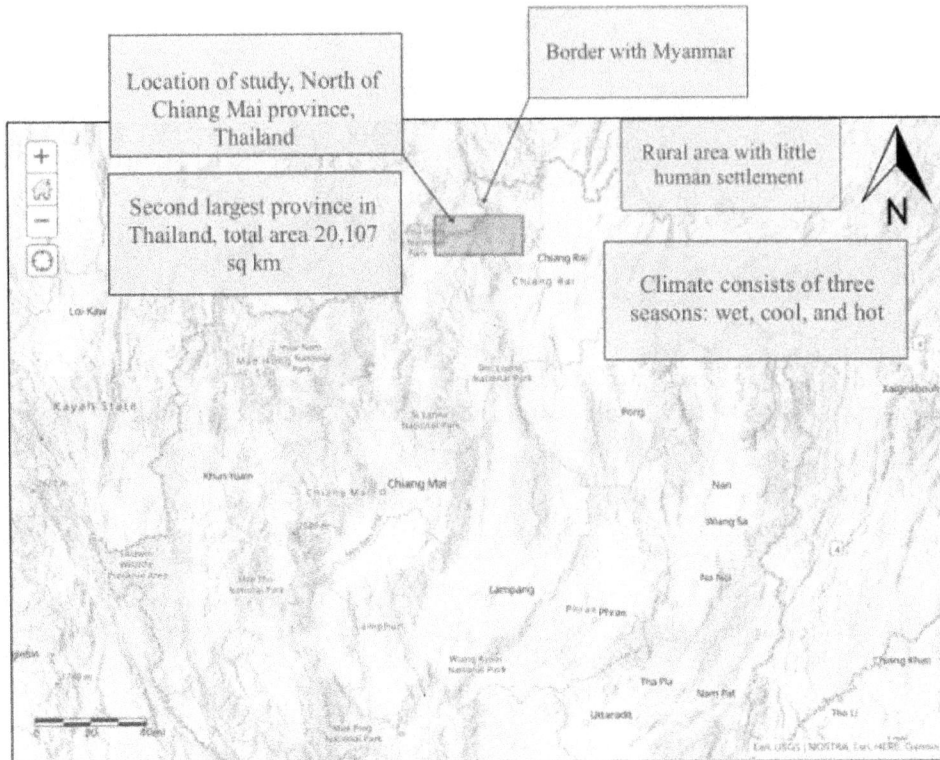

Figure 1.2: Location of study

Area of study

This river was a suitable location for the study because it allowed safe and easy access to the river sites. The river is also neither deep nor fast moving, which allows access into and across the stream. Moreover, there are numerous irrigation schemes down the river which provides the opportunity to study human impact on the river system (Figure 1.3).

Figure 1.3: Area of study

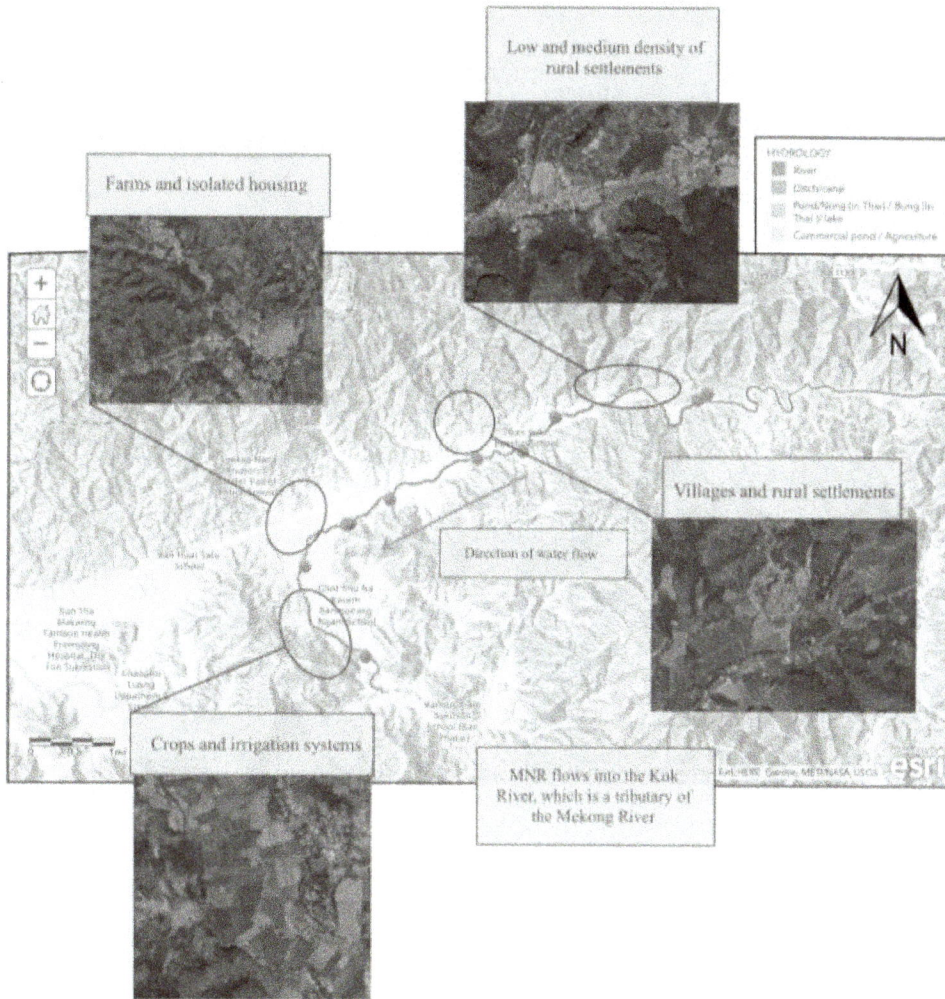

Geographical Theory

Bradshaw Model (Figure 1.4) states in theory that the discharge increases downstream of the river, due to the input from more tributaries eventually increasing the amount of river basin above that point. Also, the model predicts that the load particle size decreases downstream, as the eroded material becomes smaller due to attrition. I expect that my primary data collection will confirm the accuracy of the theoretical model, therefore my prediction is that as the discharge increases, the load particle size will decrease.

Figure 1.4: Bradshaw Model

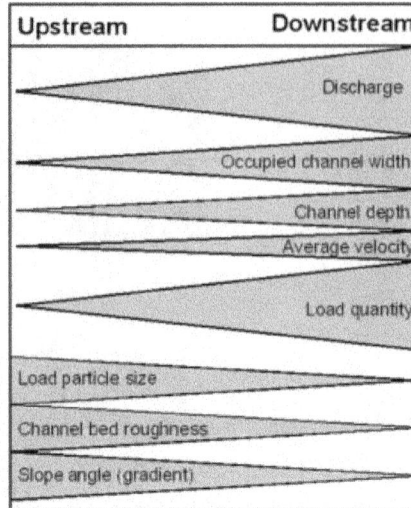

Hypotheses

Based on the secondary information, I have formed two hypotheses in my investigation:

Hypothesis 1: The sediment size of Muang Ngam River river decreases downstream
Hypothesis 2: Discharge of Muang Ngam River river increases downstream

Methods of Primary Data collection

Line random sampling was used to enable comparison between location intervals, and which allowed us to measure downstream changes in discharge and sediment size. However, due to safety and accessibility reasons, locations are not equidistant (Figure 2.1). The data was collected during a fieldwork excursion on 10-11th December 2021.

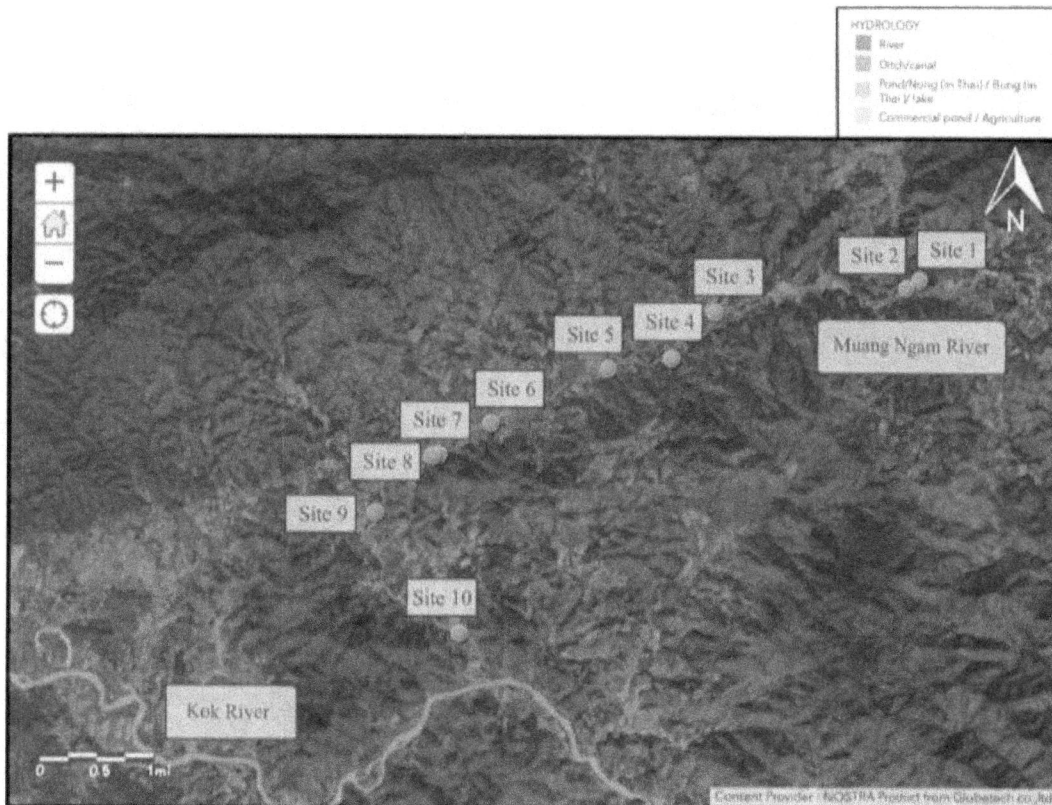

Width

We extended a tape measure from the point where the dry bank meets the water to the same point on the other side. Viewing directly above the tape, the measurement for the channel width was recorded (Figure 2.1). This method has been used because it let us utilize the tools we had available (measure tape), as well as was the simplest method to take width measurements.

Depth

In 10 regular intervals we recorded the distance between the river bed and the surface of the water, by placing a meter ruler into the river. For some sites, we used a greater number of intervals, for example 12 to 16, when the bed was more irregular, rough and bumpy. Then we calculated the mean, by adding the depths from intervals together and dividing by the number of recordings (Fig 2.2). We have decided on this method because it allows us to get an accurate representation of the average depth of the river channel, which may vary depending on its roughness.

Figure 2.2: Measuring width and depth of the river

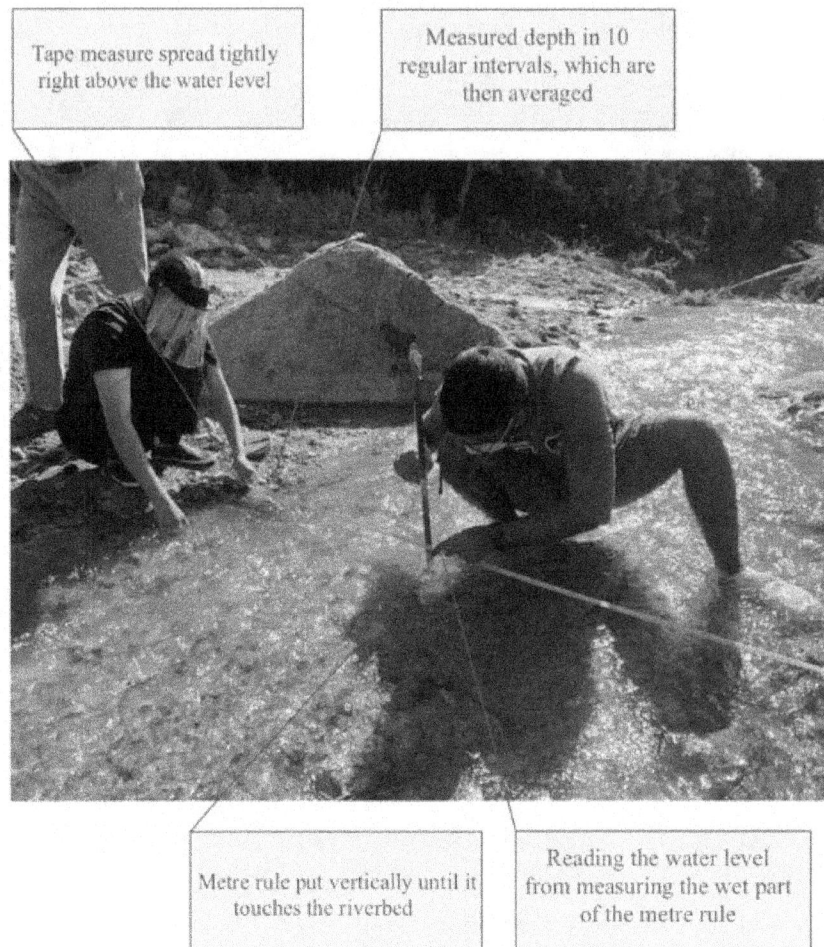

Tape measure spread tightly right above the water level

Measured depth in 10 regular intervals, which are then averaged

Metre rule put vertically until it touches the riverbed

Reading the water level from measuring the wet part of the metre rule

Velocity

Using the flowmeter method, velocity has been recorded in 3 intervals across the width of the river channel (Figure 2.3). We have decided on this method because it is a lot more accurate and time saving than for example the float method.

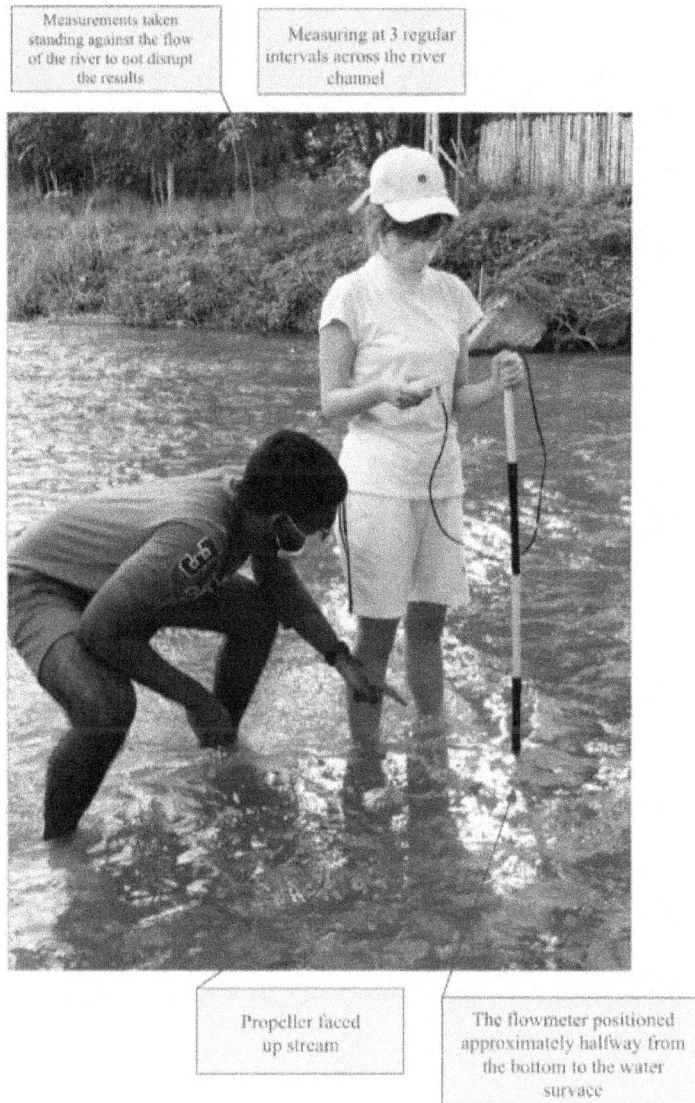

Figure 2.3: Measuring velocity

Sediment size

The pebbles have been collected with a tape measure spread tightly across the river channel. At every regular interval we have taken a sample of one pebble from the river bottom randomly (Figure 2.4). Using a vernier caliper, we measured the long, medium and short axis of each

(Figure 2.5) and compared each pebble to the rock angularity chart (Figure. 2.6). We have decided on using Power's Scale of Roundness, because it allows quick characterization of the shape of the stone and transformation of qualitative data into quantitative data.

Figure 2.4: Collection of the sediment

Figure 2.5: Measuring collected pebbles

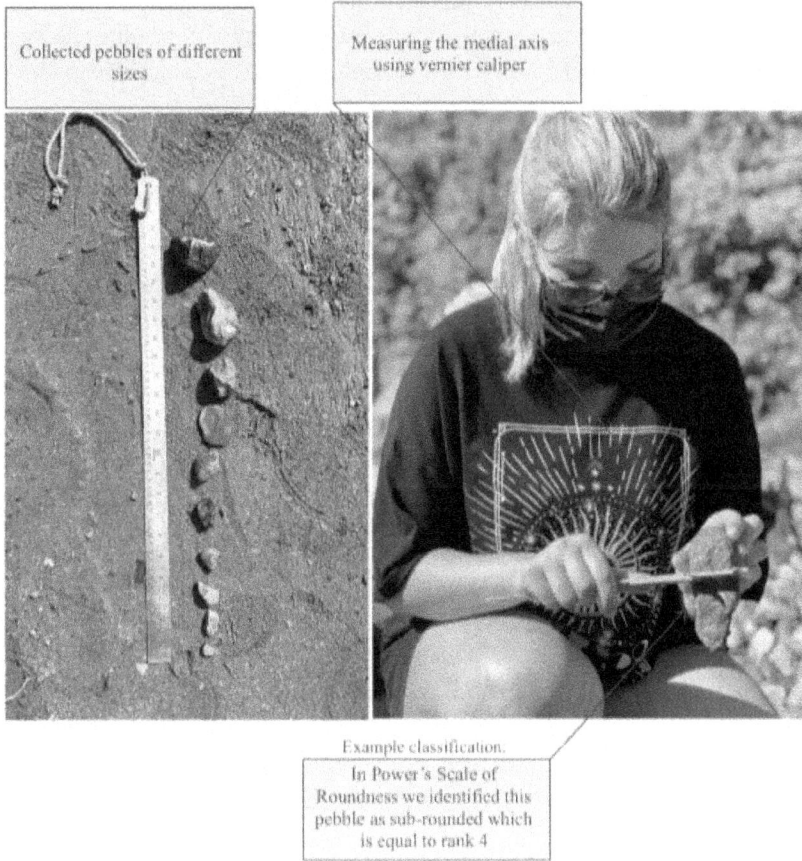

Collected pebbles of different sizes

Measuring the medial axis using vernier caliper

Example classification.
In Power's Scale of Roundness we identified this pebble as sub-rounded which is equal to rank 4

Figure 2.6: Rock angularity chart (Power's Scale of Roundness)[1]

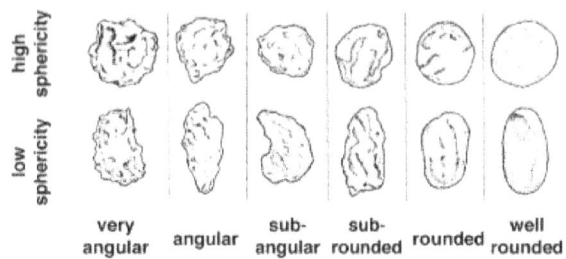

high sphericity					
low sphericity					
very angular	angular	sub-angular	sub-rounded	rounded	well rounded

---

[1] 'A new roundness scale for sedimentary particles (after Powers)' Saputra, Wardana. (2016). Problem Encountered When Producing Carbonate Sand Reservoir at researchgate.net

Methods of Secondary Data collection

To investigate additional factors which may influence downstream changes in river fluvial characteristics, I also used secondary data sources. Those include maps and satellite photographs provided and created using ArcGIS software by Esri,[2] as well as the geographical theory.

Analysis and discussion

Hypothesis 1

In order to calculate discharge using collected data of width, depth and velocity, I used the following formulas and displayed results in table 3.1:

$$Discharge \ (m^3/s) \ = \ Cross \ sectional \ area \ (m^2) \ \times \ Velocity \ (m/s)$$

$$Cross \ sectional \ area \ (m^2) \ = \ Width \ (m) \times Mean \ depth \ (m)$$

Table 3.1: Calculated values of discharge

Site	Channel width (average)	Channel depth (average)	Velocity - Flowmeter Method (average)	Discharge
1	1.23	0.0469	0.3836	0.02212873
2	1.96	0.072	0.71185	0.10045627
3	2.97	0.176	1.27	0.6638544
4	4.18	0.1765	0.818	0.60349586
5	4.96	0.123	0.83	0.5063664
6	6.1	0.15	0.932	0.85278
7	11.5	0.138	0.75	1.19025
8	6.4	0.145	1.18	1.09504
9	12.7	0.18	0.595	1.36017
10	7	0.2175	0.619	0.9424275

The data from each site was eventually plotted on a scattergraph presented in Figure 3.1. The graph suggests a steady, positive relationship between discharge of the river and distance from the source.

---

[2] "About ArcGIS | Mapping & Analytics Platform - Esri." https://www.esri.com/en-us/arcgis/about-arcgis/overview. Accessed: March 15, 2022

Figure 3.1: Scatter Graph presenting discharge at each site

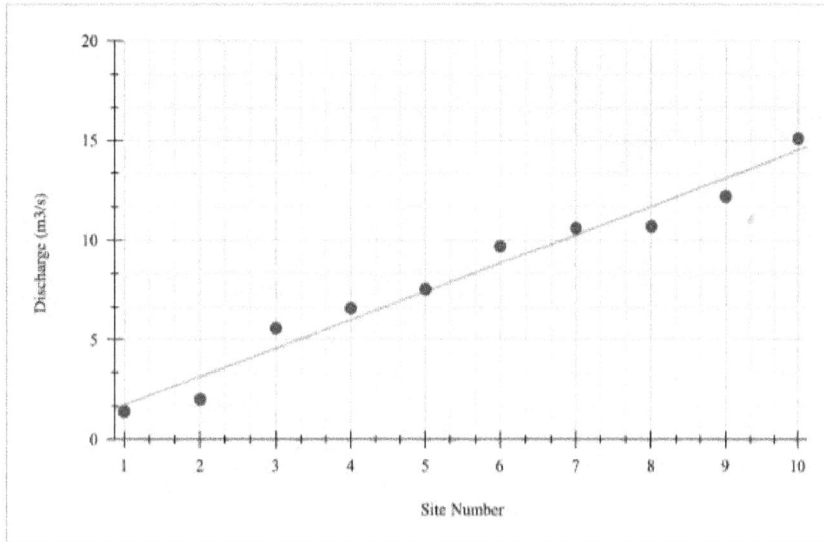

Anomalies in discharge values

There are no substantial anomalies in the trend of discharge I have observed. Yet, it is worth noticing that Site 7 to Site 9 show a decline. The map suggests that this can be explained by the significant human interference present at the sites, evidenced by straight banks of river channels. Site 8 is located below a dam (Figure 3.2), which explains a significant disruption in river's discharge. On the other hand, site 9 is located below a bridge on river channel, which causes obstacles for flood flow by reducing the cross-sectional area and inducing local eddy currents.[3] Irrigation channels before site 9 can also contribute to reduction of the river discharge downstream and increase drainage flow.

---

[3] Wang, Wen, Kaibo Zhou, Haixiao Jing, Juanli Zuo, Peng Li, and Zhanbin Li. 2019. "Effects of Bridge Piers on Flood Hazards: A Case Study on the Jialing River in China" *Water* 11, no. 6: 1181. https://doi.org/10.3390/w11061181

Figure 3.2: Possible reasons for unproportional data

Statistical Testing

H_1: Discharge of Muang Ngam River increases with increasing distance downstream.

H_0: Discharge of Muang Ngam River is independent of distance downstream.

Initial descriptive techniques have revealed a general positive trend between discharge and distance downstream (Figure 3.1), supporting my primary research hypothesis. In order to determine the strength and significance of the correlation, I used the Spearman's Rank Correlation Coefficient (SRCC). It will help me check whether the variables independently increase. I will use the significance level of 0.05, indicating that there is 0.5% chance that results occurred by chance. No anomalies or outliers were identified to be removed from the data set.

Table 3.2: Calculation for Spearman's rank correlation coefficient

Distance from source (m)	Rank (R1)	Discharge (m3/s)	Rank (R2)	Difference between Ranks (d = R1 − R2)	d2
1.38	1	0.022	1	0	0
2	2	0.1	2	0	0
5.57	3	0.66	5	-2	4
6.57	4	0.604	4	0	0
7.53	5	0.51	3	2	4
9.68	6	0.86	6	0	0
10.6	7	1.19	9	-2	4
10.7	8	1.09	8	0	0
12.2	9	1.36	10	-1	1
15.1	10	0.94	7	3	9
				Sum of d2 ($\Sigma d2$)	22
				number of pairs (n)	10
				n3-n	990
				Coefficient (Rs = 1 − ($6\Sigma d2/(n3-n)$))	0.87

SRCC showed a statistically significant result of r = 0.87, which is greater than the critical value of 0.5636 from the Spearman's rank table (Appendix 2). Since the probability of obtaining this result by chance is less than 0.05, the study successfully rejects the null hypothesis and accepts the alternative one: There is a strong positive relationship between water quality in the SMR and distance downstream.

Obtained result can be explained by the Bradshaw Model (Figure 1.4), which predicts the discharge increases downstream of the river, due to the input from more tributaries eventually increasing the amount of river basin above that point.

Hypothesis 2
After processing the data, I calculated the mean size of each pebble, which will express the sediment size. To do that, I used the following formula:

$$Mean\ pebble\ size = \frac{Long\ axis\ (cm) + Medial\ axis\ (cm) + Short\ axis(cm)}{3}$$

Calculated data (Appendix 1) has been eventually plotted on a diagram on a map presented in Figure 3.3, which describes mean pebble size at every site. Despite a very accurate visualization, the map does not suggest a clear trend in the sediment size downstream. Therefore, I decided to construct another scatter plot (Figure 3.4) and a trendline, which would indicate either a positive or negative relationship.

Figure 3.3 Map presenting visualized pebble size at each site

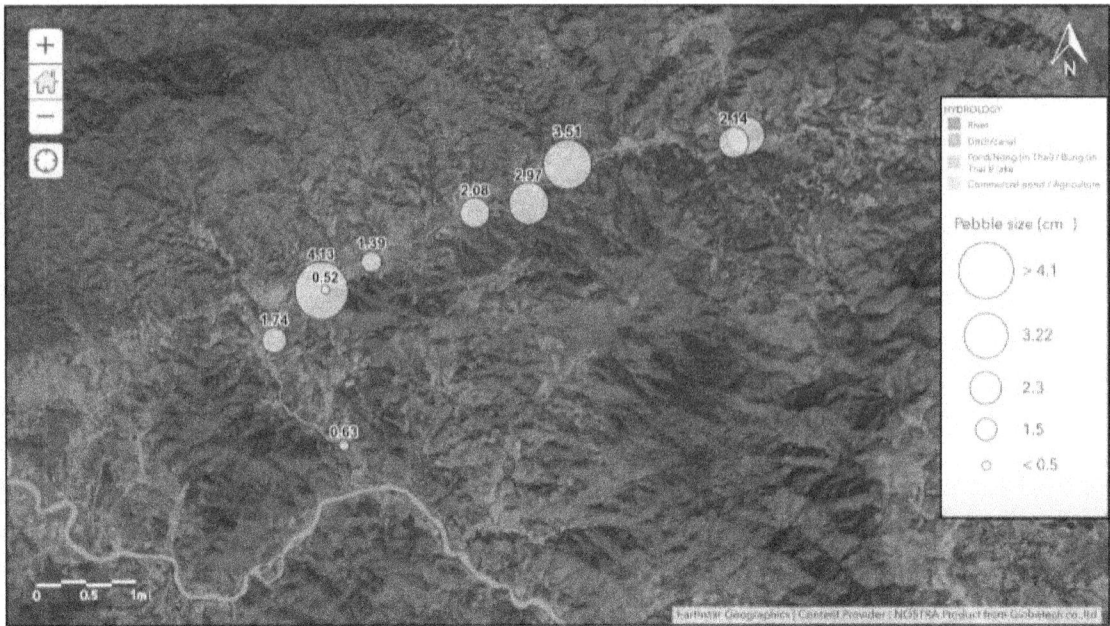

Figure 3.4: Scatter graph demonstrating mean pebble size sampled from each site

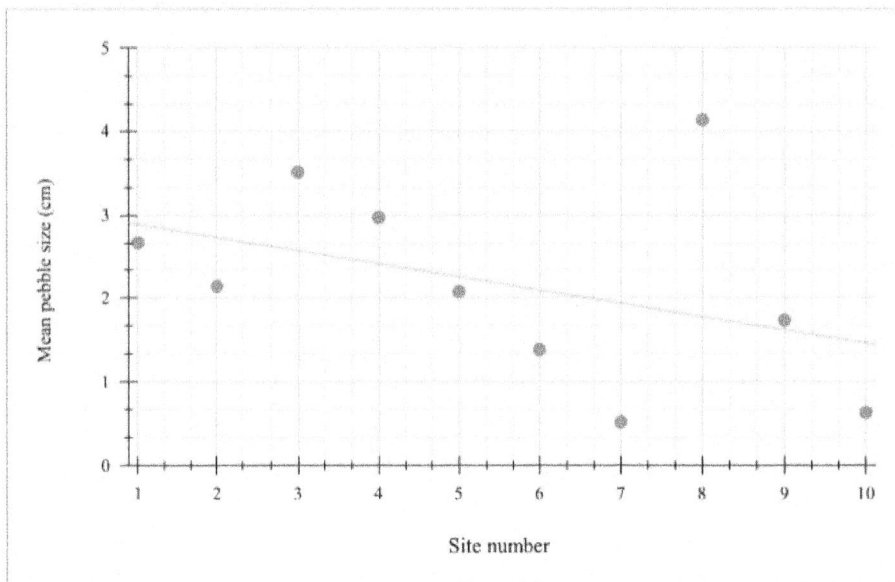

Anomalies in pebble size

From figure 3.4, I observed two anomalies in the sediment size values, which are far away from the trendline. I identified these anomalies as values from site 7 and site 8. Also, it is worth noticing that value from site 3 is a lot greater than the average and far from the trendline.

I deduced that the anomaly at site 7 is a result of the unique conditions of that area. The area of site 7 was previously a reservoir, the river at this site had very fine sediments and muddy water (Figure 3.5). It is because of the heavy rainfall, which has deposited sediments due to erosion uphill behind the dam. Moreover, sediments which were dug out from the steep banks had fallen into the river.

On the other hand, I suggest that anomaly at site 8, is a result of a methodology of data collection. I believe that one of the rocks, which is multiple times bigger than the rest (Figure 3.5), has skewed the average size of all the pebbles. Site 8 is below a dam, therefore this anomaly is not supported by geographical theory. It would suggest that the biggest sediments would get built up by the dam, leaving only small pebbles below. Thus, I deduce that these anomalies were a result of flawed methodology. Sampling more than one pebble from each interval across the river channel would ensure greater accuracy and less bias.

Similarly to site 8, I believe the above average pebble size from site 3 was also a result of flawed data collection. I predict that even though the pebbles were supposed to be sampled randomly, a person sampling might have had a natural preference towards grabbing easily accessible rocks of bigger size, which explain this significantly high mean size of pebbles at Site 3.

Figure 3.5: Sediment collected at Site 3, site 7and site 8

Statistical testing

H_1: Sediment size of Muang Ngam River decreases with increasing distance downstream.
H_0: Sediment size of Muang Ngam River is independent of distance downstream.

Initial descriptive techniques and graph trend lines have shown a general negative trend between sediment size and distance downstream (Figure 3.3), which supports my alternative research hypothesis. Although I identified 3 different values of mean pebble size as anomalies, none of them is an outlier exceeding the upper boundary of data. Therefore, I decided not to exclude them from the dataset and statistical testing (Appendix 3).

SRCC has been again used and showed a statistically significant result of $r = -0.47$, which is greater than the critical value of 0.4424. Since the probability of obtaining this result by chance is less than 0.1, the study successfully rejects the null hypothesis and accepts the alternative one: Sediment size of Muang Ngam River decreases with increasing distance downstream. It is important to note that although the negative correlation was of moderate strength, the relationship does not suggest causation between the two variables, there can also be external factors responsible for the relationship. Nevertheless, such SRCC findings reveal the interdependent relationship between sediment size of MNR and distance downstream.

The negative relationship can be explained by initial secondary research, and Bradshaw model (Figure 1.4) which predicts why the sediment size decreases downstream. It happens because the further downstream material is carried, then the greater the time available for it to be eroded by attrition and abrasion which makes rocks and stones smaller and rounder. Also, when discharge is high vertical erosion erodes the river bed and larger sediments are transported by traction. Therefore the sediment load of the river gets smaller in size.[4]

To further support my hypothesis, I also analyzed the change in shape of the river sediment downstream. The graph in Figure 3.3 suggests that further downstream, the shape of the pebbles increased in the Power's Scale of Roundness, meaning that with each site they became less angular and more rounded. That can also support my alternative hypothesis, suggesting that the attrition forces not only decreased the size of the sediment, but also its shape.

---

[4] 'Why does load size decrease downstream?' Fleming Esther. Last accessed March 15, 2022: https://www.sidmartinbio.org/why-does-load-size-decrease-downstream/

Figure 3.6: Line graph presenting the pebble shape in Power's Scale of Roundness at each site

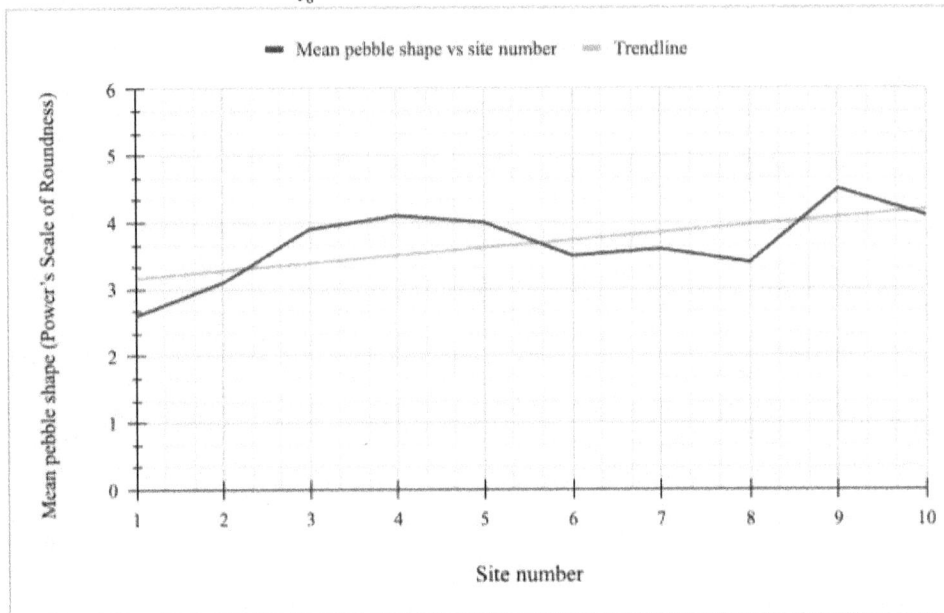

Conclusion

This study helped me successfully answer the research question "How do the fluvial characteristics of Muang Ngam River, Chiang Mai, change with distance from the source?". I was able to conclude a number of points:

Hypothesis 1

At each site downstream, the value of discharge increased forming a steady and proportional trend (Figure 3.1). In accordance with my hypothesis, the study found a strong and statistically significant positive correlation between discharge and distance downstream. This is consistent with The Bradshaw Model, which states that in theory, the discharge increases downstream of the river, due to the input from more tributaries eventually increasing the amount of river basin above that point.

Hypothesis 2

The study determined a moderately strong negative correlation between the average sediment size and distance downstream. At each site downstream, the size of pebbles became generally smaller (Figure 3.3) and the shape rounder (Figure 3.5). In accordance with my predictions and hypothesis, sediment size of Muang Ngam River decreases with increasing distance downstream. It is also consistent with The Bradshaw Model, which states that in theory, that the further material is moved downstream, the more time it has to be eroded by attrition and abrasion, which reduces the size and rounds the shape of pebbles and stones.

Evaluation and improvements

Limitations	Impact	Method of Improvement
The study does not account for other geographical factors	Any rainfall prior to data collection would affect the discharge of the river, which could affect the displacement of sediment thus influencing the reliability of the data.	Any rainfall a few days prior to the data collection should be noted. This allows us to account for increased discharge due to rainfall thus increasing the reliability of the study
The study does not account for all human factors	The presence of land of agricultural use, crops, irrigation channels, bridges and dams heavily influences the fluvial characteristics and dynamics of the river.	A detailed observation of every human interference that could potentially affect the cross-sectional area or velocity of the whole course of the river should be provided. This allows us to account for disturbed discharge due to human factors and increases the reliability of the study.
Not equidistant site locations	The lack of equidistant locations greatly diminishes data accuracy as we could only collect limited data from specific locations. The limited data is likely to be unreliable and does not provide us with a holistic understanding of the entire river.	Data should ideally be collected at evenly spaced-out sites, allowing us to improve the accuracy of our results and make stronger generalizations.
Small sample size	The sample size was not big enough to provide a comprehensive overview of the size of the sediment size.	The number of pebbles collected from every interval should be a lot bigger (10-20), which would help us gain more understanding of the possible trend and decrease the possibility of significant anomalies in data.

BIBLIOGRAPHY + APPENDICES OMITTED

2. EXAMPLE TWO (23/25)

Title: 'What Socio-Economic Impacts Does Tourism Bring to Rural Communities

in The Mae Ai Region Of Thailand?'

Author: Anonymous

Session: May 2022

Level: HL

Examiner's summary

Criterion A [3/3]:

The student has successfully formulated a narrowly focused geographical fieldwork question and provided a clear location for the study. They have also justified the choice of fieldwork location by explaining its relevance to the topic of tourism. The student has demonstrated a good understanding of geographical theory and its application to the fieldwork, as well as providing detailed predictions and hypotheses. Additionally, the student has linked the study to the geography syllabus and included well-designed locational maps with all necessary components.

Criterion B [2/3]:

The student demonstrated a clear understanding of the methods used for both primary and secondary data collection, and provided a justification for each method. They also selected appropriate techniques for collecting quality data for subsequent analysis. Although the student did not state the date and time of data collection, they did provide the exact location. Additionally, all necessary figures were included to explain the methods of the investigation.

Criterion C [6/6]:

The student has collected and presented a sufficient amount of relevant data for the fieldwork question, including tables with data, graphs, photos, and maps. All figures used by the student are appropriately labeled with a title, cardinal directions, and a scale. The sample size used for analysis is detailed and in-depth, consisting of many primary and secondary data. The student used a variety of different techniques, including well-annotated and personalized maps, graphs, tables, and photos to present the collected data.

Criterion D [8/8]:

The student provided a thorough and detailed discussion about the collected data and correctly interpreted the chi test results. They also recognized and analyzed significant trends and patterns in the data and identified important outliers/anomalies and suggested their potential source. The descriptive techniques and applied statistical tests were appropriately selected for the data collected and the fieldwork question. Additionally, the written analysis is relevant to the posed fieldwork question and assumed predictions/hypotheses, and the student referred to geographical context and theory.

Criterion E [2/2]:

The student's conclusion clearly answers the fieldwork question and is supported by the data collected and its analysis. They referred to their original predictions/hypotheses and compared them with the findings, and summarized the results of the fieldwork investigation. Additionally, the student included the most important data and analysis to support their conclusion.

Criterion F [2/3]:

The student effectively identified the weaknesses of the fieldwork methodology and correctly recognized potential factors that could have affected the reliability of the collected data. However, the student did not state any strengths of the methodology. The student suggested specific and doable improvements to enhance the methodology, which were well-explained and justified.

Other requirements [0/0]:

The student has met the criteria for the specified criterion as they have not exceeded the word limit of 2500 words, numbered their pages correctly, and included references to all external sources of information.

Geography Internal Assessment

Option E - Leisure, Sports and Tourism

What Socio-Economic Impacts Does Tourism Bring To Rural Communities in The Mae Ai Region Of Thailand?

Figure 1- Wat Thaton temple in the Mae Ai district

Chang (2016)

glt506

Word count- 2457 words

Introduction- 216 words

Methodology- 308 words

Quality and Treatment of data- 1468 words

Conclusion- 132 words

Evaluation- 333 words

Introduction

Fieldwork question: What socio-economic changes does tourism bring to rural communities in the Mae Ai region of Thailand?

Geographical context:

Figure 2- A Peter's projection of the world map

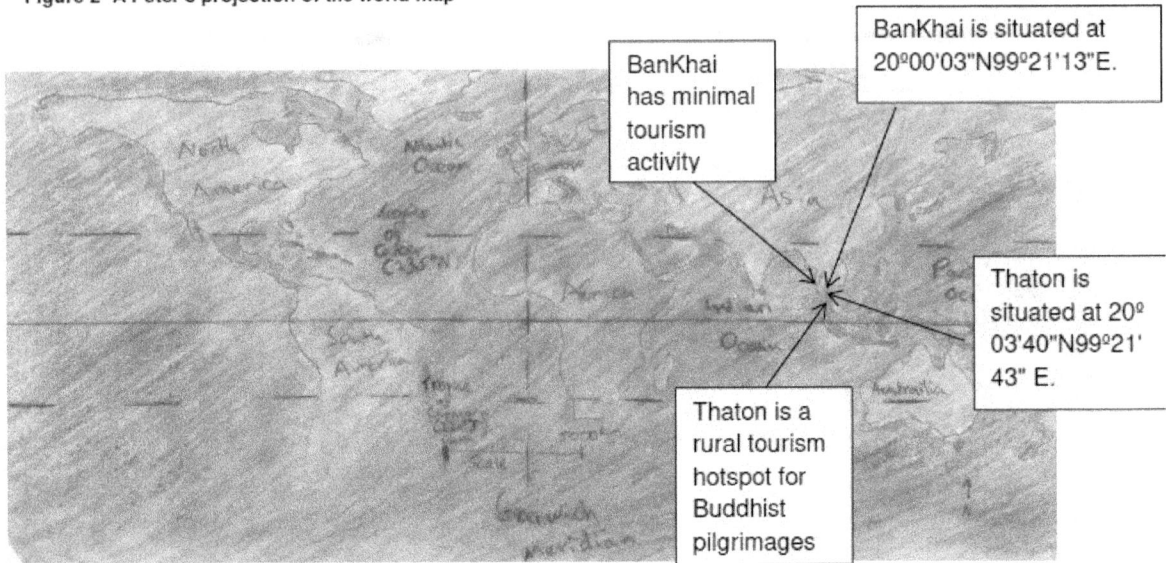

BanKhai has minimal tourism activity

BanKhai is situated at 20º00'03"N99º21'13"E.

Thaton is situated at 20º 03'40"N99º21' 43" E.

Thaton is a rural tourism hotspot for Buddhist pilgrimages

(Hydrosheds , 2010)

Figure 3- Regional map of the Thailand showing Mae Ai

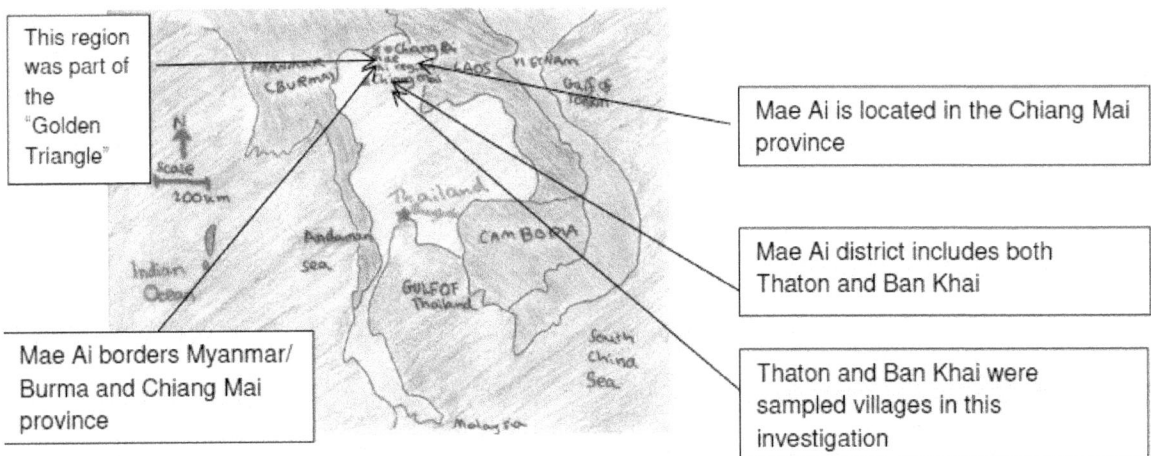

This region was part of the "Golden Triangle"

Mae Ai is located in the Chiang Mai province

Mae Ai district includes both Thaton and Ban Khai

Mae Ai borders Myanmar/ Burma and Chiang Mai province

Thaton and Ban Khai were sampled villages in this investigation

(Lonely planet, 2014)

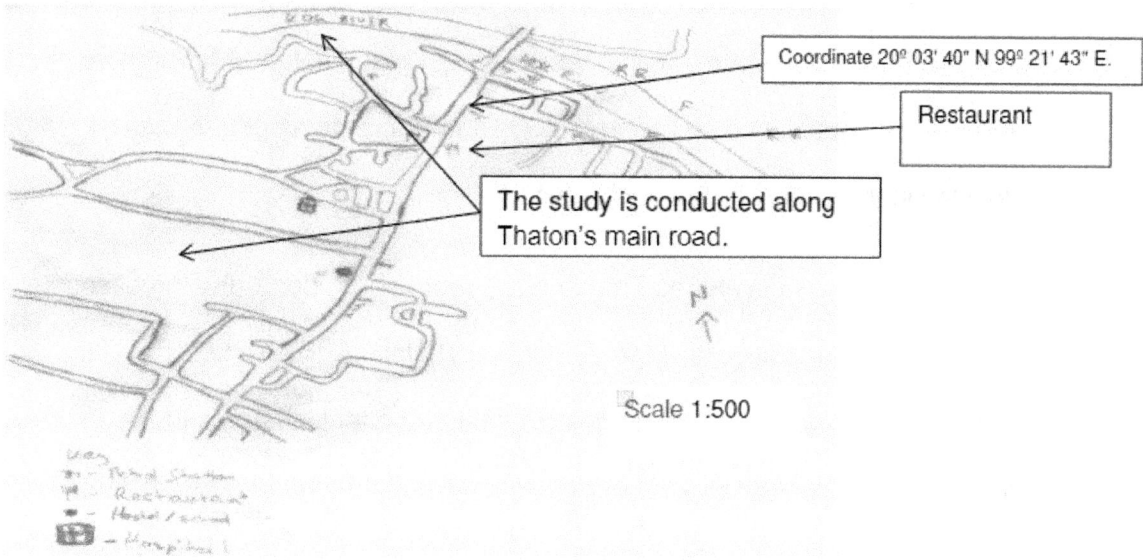

Coordinate 20º 03' 40" N 99º 21' 43" E.

Restaurant

The study is conducted along Thaton's main road.

N

Scale 1:500

(motionX, 2016)

Figure 5- Regional map of BanKhai, a region with minimal tourism activity (similar situation to Thaton 20 years ago)

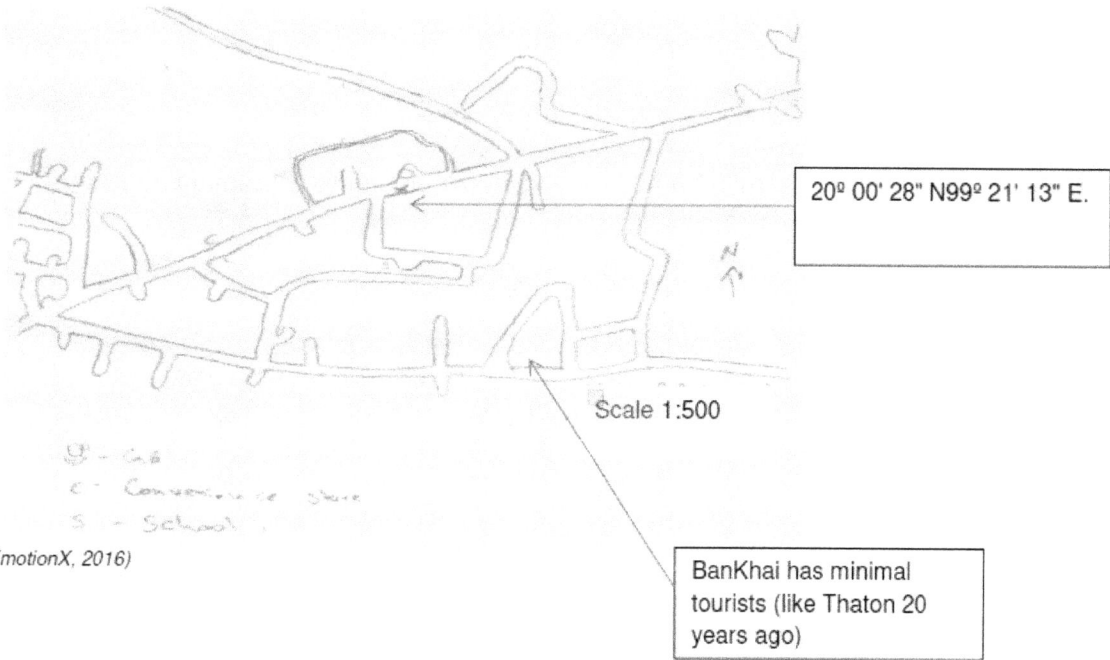

20º 00' 28" N99º 21' 13" E.

N

Scale 1:500

(motionX, 2016)

BanKhai has minimal tourists (like Thaton 20 years ago)

Socio-economic context (Thaton and BanKhai):

Thaton is a domestic tourism hotspot. Thaton is a honeypot site for Buddhists completing a pilgrimage due to its many temples e.g. Wat Thaton (Massington, 2016).Thaton is a nodal point of the Chiang Mai/ Chiang Rai route along the Maekok River. It is located in the Mae Ai region of Chiang Mai.

BanKhai resembles the less developed Thaton of 20 years ago (prior to the increase in tourism) and is another village in Mae Ai. Unlike Thaton (which is tourism), its main income source is agriculture. Tourism is minimal and oftentimes people live in larger families, and development is gradually commencing in the region (Massington, 2016).

Therefore, this investigation seeks to compare the socio-economic impacts of tourism on the region of Mae Ai by comparing a tourism hotspot with an area with minimal tourism.

Syllabus link: The fieldwork links to Option E- Leisure Sports and Tourism and specifically to the point: Examine the economic, social and environmental impacts of tourism on a local scale (IBO, 2009).

Hypothesis and theory: *Tourism brings Mae Ai mostly positive socio-economic impacts.* This is because in accordance with the Multiplier-effect and Butler's model (reference figures 6&7) tourism spending will stimulate development which will improve living standards and the multitude tourist interactions will have positive social benefits.

Myrdal's Multiplier Effect

Figure 6 - Myrdal's Multiplier effect

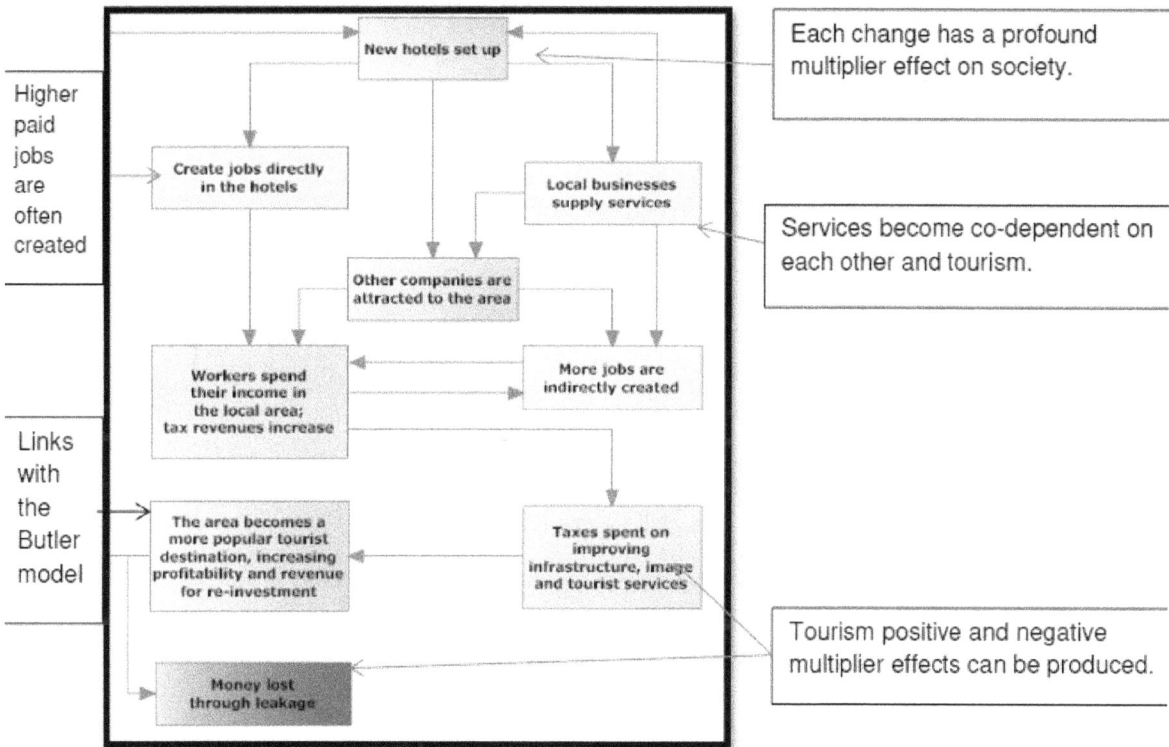

Higher
paid
jobs
are
often
created

Links
with
the
Butler
model

New hotels set up

Create jobs directly
in the hotels

Local businesses
supply services

Other companies are
attracted to the area

Workers spend
their income in
the local area;
tax revenues increase

More jobs are
indirectly created

The area becomes a
more popular tourist
destination, increasing
profitability and revenue
for re-investment

Taxes spent on
improving
infrastructure, image
and tourist services

Money lost
through leakage

Each change has a profound
multiplier effect on society.

Services become co-dependent on
each other and tourism.

Tourism positive and negative
multiplier effects can be produced.

(Greenfieldgeography, 2016)

63

Multiplier effect

Figure 7- Butler Model representing the relationship between
tourism and an area over time

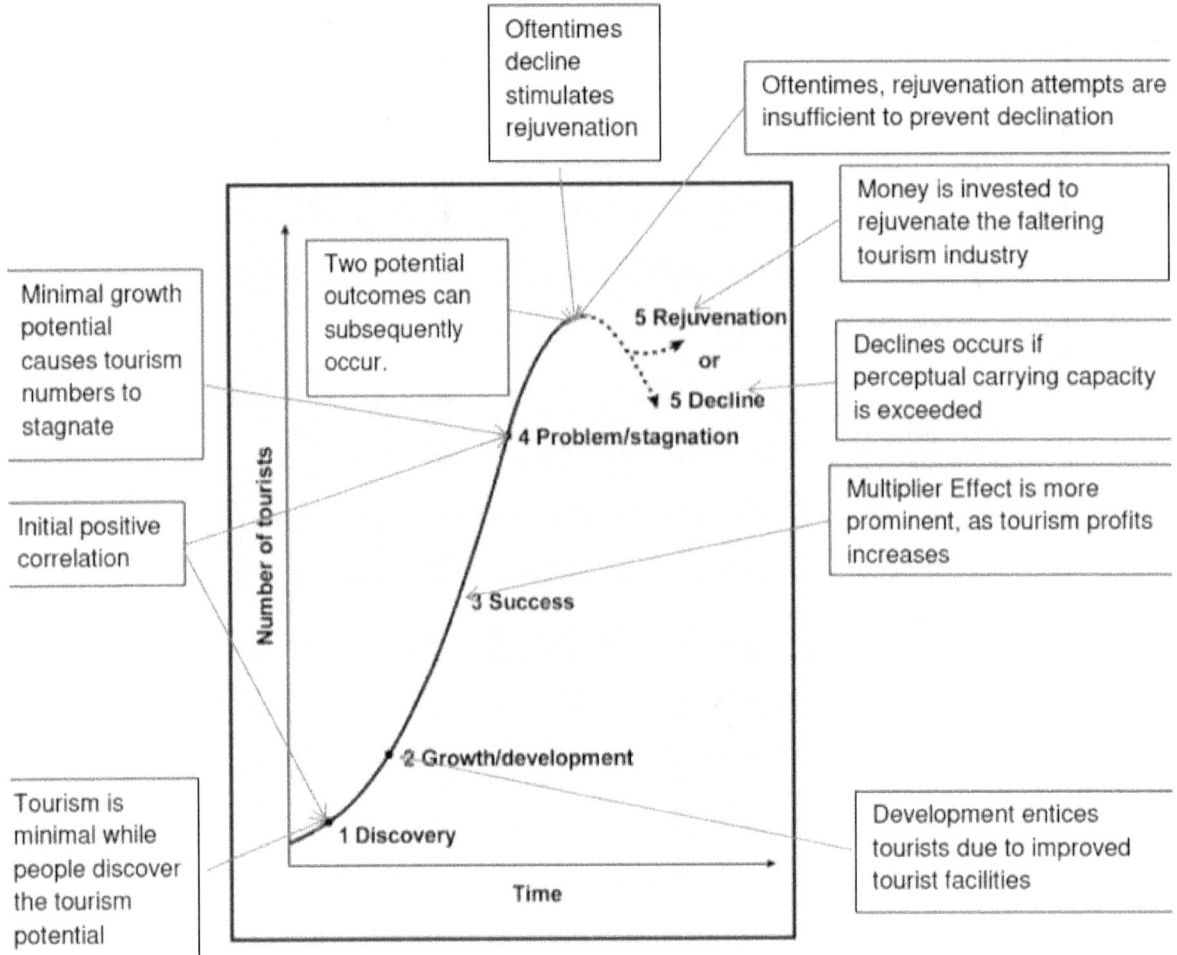

Oftentimes
decline
stimulates
rejuvenation

Oftentimes, rejuvenation attempts are
insufficient to prevent declination

Money is invested to
rejuvenate the faltering
tourism industry

Minimal growth
potential
causes tourism
numbers to
stagnate

Two potential
outcomes can
subsequently
occur.

5 Rejuvenation

or

5 Decline

Declines occurs if
perceptual carrying capacity
is exceeded

4 Problem/stagnation

Multiplier Effect is more
prominent, as tourism profits
increases

Initial positive
correlation

Number of tourists

3 Success

2 Growth/development

Tourism is
minimal while
people discover
the tourism
potential

1 Discovery

Development entices
tourists due to improved
tourist facilities

Time

(Greenfieldgeography, 2013)

Methodology

Risk assessment:

1. Wear sunblock to prevent strokes and sunburns

2. Drink 7L of water /day to prevent dehydration in tropical climates

3. Apply mosquito repellent to prevent malaria contraction

Justification of apparatus

1. Stopwatch - to ensure vehicle and pedestrian counts do not exceed 2 minutes.

2. Random number table - to prevent selection bias at sampling locations for all random samples

Secondary data rationale

Due to Mae Ai's relative obscurity, a lot of data was gathered from two British Geographers who had resided in the region for a substantial duration (although IB geography Fieldwork book often further substantiated their data). Additionally, geographical journals and textbooks such as Geographical Enquires by Nagle & Spencer were read in order to gain a better understanding of tourism concepts and how these concepts may be applicable to Mae Ai.

Selecting sites

Figure 8- Street map of Thaton showing sites sampled

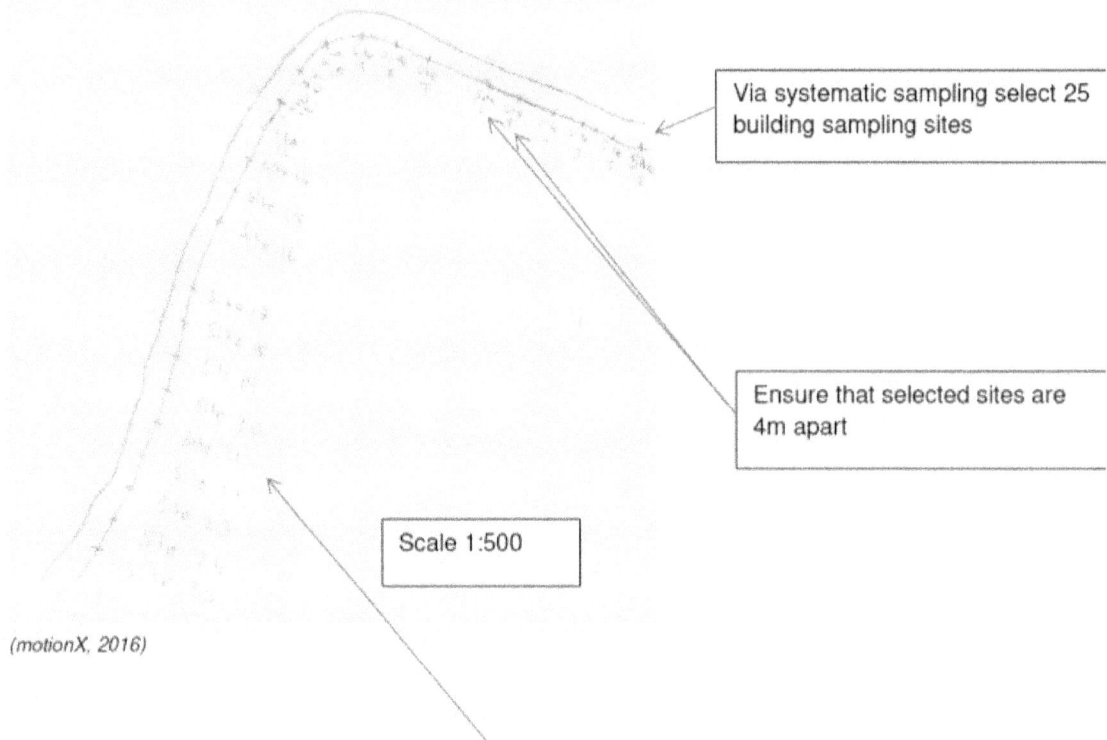

Via systematic sampling select 25 building sampling sites

Ensure that selected sites are 4m apart

Scale 1:500

(motionX, 2016)

Figure 9- Street map of BanKhai showing sites sampled

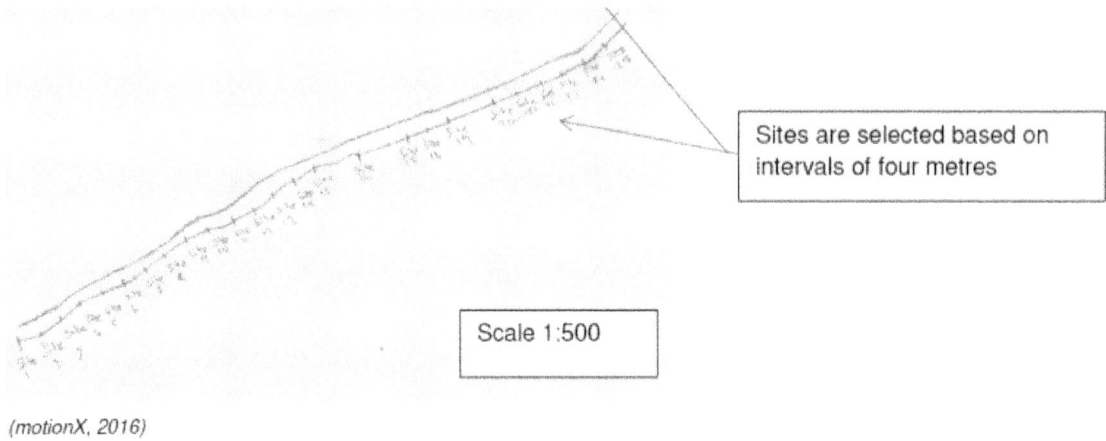

Sites are selected based on intervals of four metres

Scale 1:500

(motionX, 2016)

Determining tourism impacts on land usages

Figure 10- Photograph to show a potential site whereby a land usage survey can be conducted

Categorise buildings at 25 selected sites (reference table 8).

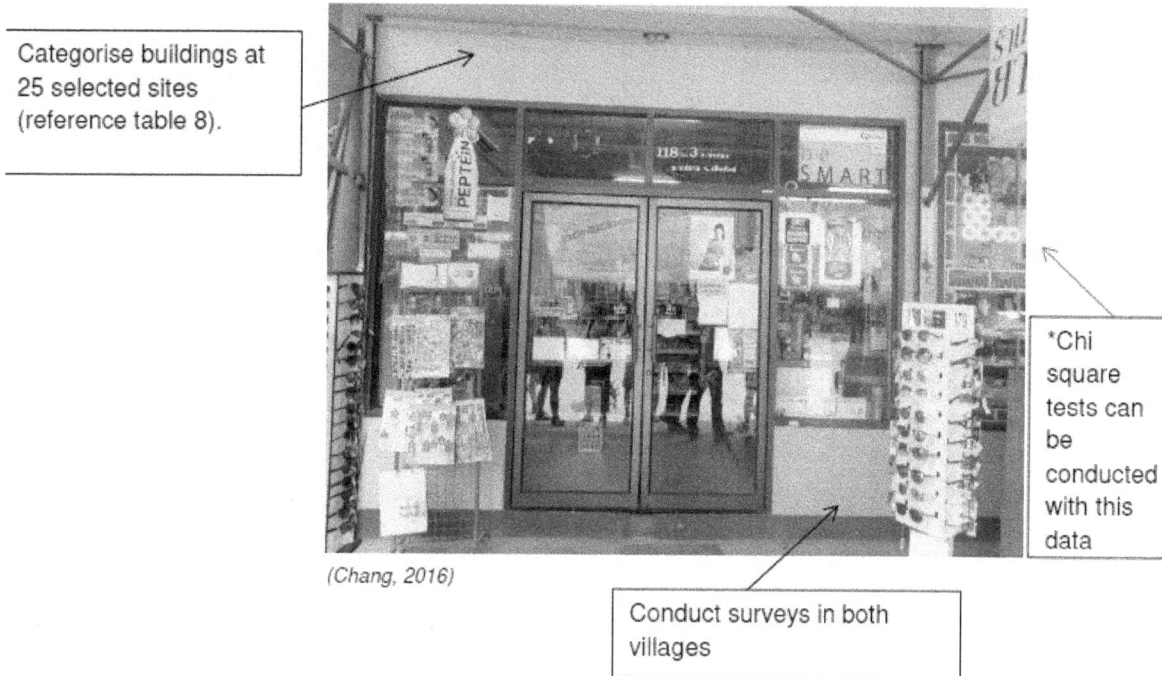

(Chang, 2016)

*Chi square tests can be conducted with this data

Conduct surveys in both villages

Justification- Land usage is a proxy indicator for a locations economic situation (i.e. more shops could mean greater economic activity). Prevalence of facilities such as massage parlours would also indicate that land usage is catered towards tourists. Additionally, if more high order goods were present, socio-economic impacts of tourism would be considered positive.

Determining tourism impacts based on pedestrian and vehicle count

Figure 11- Gathering pedestrian and vehicle counts

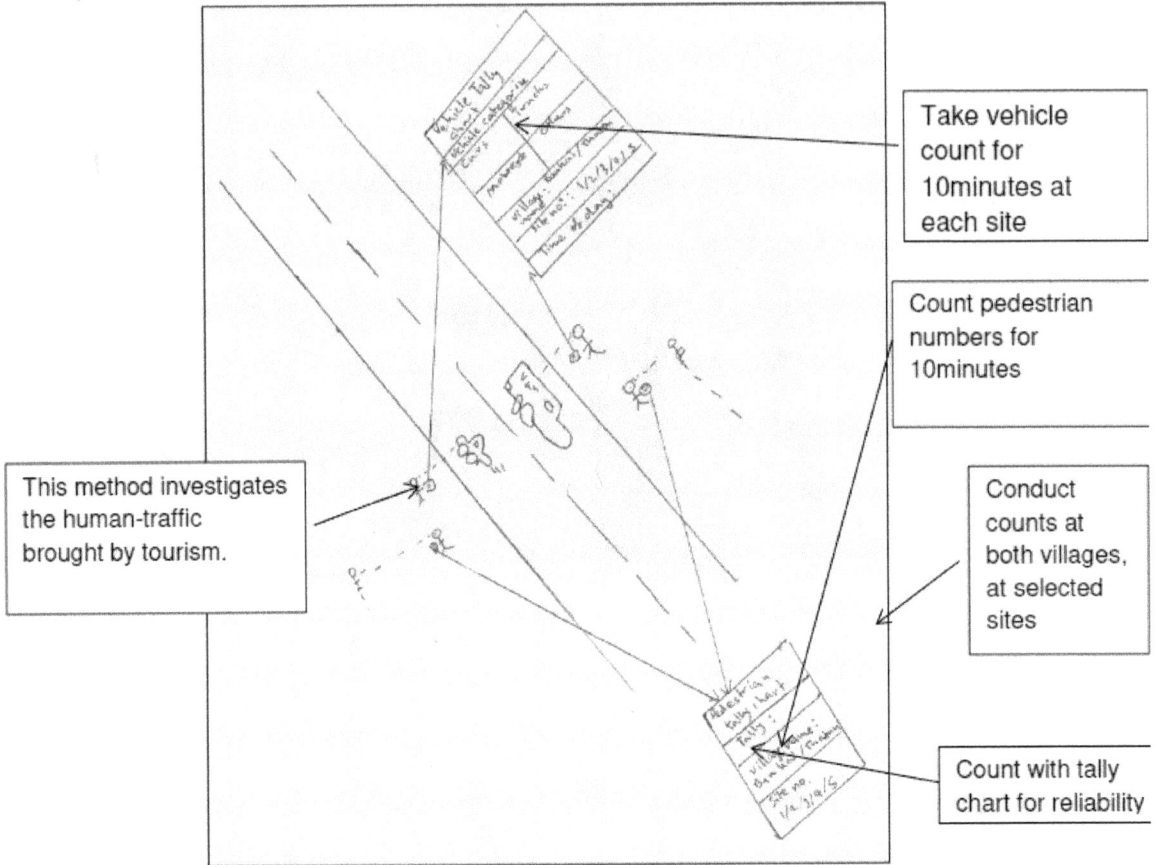

Take vehicle count for 10minutes at each site

Count pedestrian numbers for 10minutes

This method investigates the human-traffic brought by tourism.

Conduct counts at both villages, at selected sites

Count with tally chart for reliability

Justification –These counts indicate tourism volume and local spending on high-order goods such as vehicles; these are all impacts that tourism can bring to a location and can reveal disposable income and economic activity.

Determining tourism impacts based on local businesses

Only conducted in Thaton.

BanKhai is assumed to have minimal tourism.

Table 1- Local questionnaire about tourism

Via random sampling, distribute questionnaires to 6 businesses

Conduct questionnaires by asking questions to participants

Random sampling methods minimizes human partiality

Questionnaire results determine socio-economic impact on the villages businesses

Type of business:_____

1. How long has the business been operating?

2. Did you set up the business mainly for tourists?

3. What percentage of your business comes from tourism?

4. How many people (including yourself) do you employ?

5. Has the amount of business increased, decreased or stayed the same over the past few years?

6. What problems are there of relying on tourism....

Asking business managers to complete questionnaire

Translate into local language for better understanding

Table 2- Translated questionnaire (Thai)

Figure 12- Photograph of local businesswomen filling out a questionnaire

(Chang, 2016)

Justification- To analyse tourism impacts on the business economic circumstances.

Determining tourism impacts on individual perceptions

Table 3- Table containing questions regarding local perception of tourism

		Yes	No	Not sure
1	Do you like tourists visiting Thaton?			
2	Has tourism provided more jobs?			
3	Has tourism increased incomes?			
4	Do you think tourism has caused prices of goods to increase?			
5	Do you think tourism has caused the price of land to increase?			
6	Have you noticed an increase in these problems because of tourism? a) Drugs b) Drunkenness c) Crime, such as theft			
7.	Do you think the local culture has changed because of tourism?			

Ask simple questions to ensure credible answers.

Using random sampling obtain 6 questionnaire participants

Only conduct in tourist dense locations (otherwise redundant)

Figure 13- Photograph of local individual filling in questionnaire

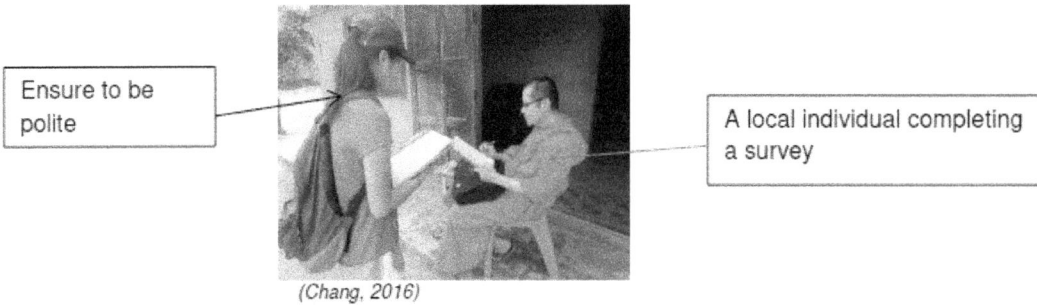

Ensure to be polite

A local individual completing a survey

(Chang, 2016)

Justification- Results gathered can determine people's perception of the tourism industry.

Determining tourism impacts on living standards

1. Via random number table select 50 buildings from 25 sites (ref. figs 9&10)

2. Rank 50 houses between 1 and 3 in each category. Reference table (below)

3. Repeat in Ban Khai

4. Justification- Provides a proxy indicator of economic impacts based on the residences accommodation standards.

Table 4- Raw data table regarding housing living conditions

Thaton/ Ban Khai	Roof material			Wall material			General Condition			Total
House	3	2	1	3	2	1	3	2	1	
1										
2										
3										
4										
5										
6										
7										
8										
9										
10										
								Total		

Measuring quality of life

Table 5- Measuring quality of life with various factors

QUALITY OF LIFE COMPARISON		VILLAGE: THA TON				VILLAGE: BAN KHAI			
		0	1	2	3	0	1	2	3
HOUSING	Quality of construction	V. POOR	POOR	SATIS	GOOD	V. POOR	POOR	SATIS	GOOD
	Roof material	90% THATCH	50% THATCH	75% TILE	90% TILE	90% THATCH	50% THATCH	75% TILE	90% TILE
	Mains electricity to houses	NONE	SOME	MOST	ALL	NONE	SOME	MOST	ALL
	SUB-SCORE								
SERVICES	Evidence of piped water	NONE	LIMITED	DEVELOPED	WELL DEV.	NONE	LIMITED	DEVELOPED	WELL DEV.
	Electricity	NONE	LIMITED	DEVELOPED	WELL DEV.	NONE	LIMITED	DEVELOPED	WELL DEV.
	Evidence of sewage disposal	NONE	LIMITED	DEVELOPED	WELL DEV.	NONE	LIMITED	DEVELOPED	WELL DEV.
	SUB-SCORE								
TRANSPORT AND COMMUNICATIONS	Quality of roads	V. POOR	POOR	SATIS	GOOD	V. POOR	POOR	SATIS	GOOD
	Variety of transport	V. POOR	POOR	SATIS	GOOD	V. POOR	POOR	SATIS	GOOD
	Telephone lines	NONE	VERY FEW	SOME	MANY	NONE	VERY FEW	SOME	MANY
	SUB-SCORE								
EVIDENCE OF DISPOSABLE INCOME	Number of shops / services	NONE	LIMITED	SOME	MANY	NONE	LIMITED	SOME	MANY
	Quality and range	NONE	LIMITED	SATIS.	GOOD	NONE	LIMITED	SATIS.	GOOD
	ATM Machines	NONE	1	2	3	NONE	1	2	3
	Motor bikes	NONE	VERY FEW	SOME	MANY	NONE	VERY FEW	SOME	MANY
	Cars / Pick-up trucks	NONE	VERY FEW	SOME	MANY	NONE	VERY FEW	SOME	MANY
	SUB-SCORE								
EDUCATION	Condition of the classrooms	V. POOR	POOR	SATIS	GOOD	V. POOR	POOR	SATIS	GOOD
	Condition of the canteen	V. POOR	POOR	SATIS	GOOD	V. POOR	POOR	SATIS	GOOD
	Condition of the library	V. POOR	POOR	SATIS	GOOD	V. POOR	POOR	SATIS	GOOD
	Condition of the sports field	V. POOR	POOR	SATIS	GOOD	V. POOR	POOR	SATIS	GOOD
	General impression of students	MOST POORLY DRESSED			MOSTLY WELL DRESSED	MOST POORLY DRESSED			MOSTLY WELL DRESSED
	TOTAL SCORE								

Survey sites referenced in figures 4 and 5

0 to 3 is in improving order

Have 6 groups of 10 complete this

Total up average scores

As participant numbers increases, human biasness decreases

How to quantify something based upon description

Specific detail of each category

Justification- To determine if quality of life improves with increased tourist quantities.

Price comparison

Table 6- Price comparisons of goods between two locations

OTHER COMPARISONS	(One rai)			
1. Price of land	a.			
2. Price of goods	b.			
	c.			
	d.			
	e.			
	f.			

Select 5 items to show any inflation

Secondary data required

Compare the price of identical products at two sites

Justification- Determining if tourism causes inflation in the Mae Ai district.

73

Quality and treatment of data

Trends in land usages in both villages:

Table 7 - Land usages in Thaton and BanKhai

BanKhai has 8 times more manufacturing (secondary industry) services.

Reference table 8 page 19 for key

High-order goods are more plentiful in Thaton

5 more restaurants (E2) in Thaton than BanKhai

	BanKhai		Thaton
R1	3	R1	6
R2	6	R2	6
I1	4	I1	0
I2	4	I2	1
C1	1	C1	1
C2	0	C2	1
C3	0	C3	0
C4	0	C4	0
C5	4	C5	9
C6	1	C6	2
C7	3	C7	5
P1	1	P1	0
P2	1	P2	1
P4	0	P4	1
P5	0	P5	1
E1	1	E1	1
E2	2	E2	7
E3	0	E3	7
O1	0	O1	0
O2	0	O2	0
O3	0	O3	0
O4	4	O4	0
U1	5	U1	2
U2	3	U2	3
U3	0	U3	1

Table 7 (above) shows the differences between the land usages in the two villages. Thaton's land use has 8 examples of high order services e.g. shops, ATMs, car rentals etc (as opposed to 5 in BanKhai). Additionally there are 5 more restaurants in Thaton which suggests that facilities in Thaton are catered towards tourists or that Thaton residences have greater disposable incomes (Ozgener, et al, 2006) as theorised by Myrdal's multiplier effect. Such facilities earn money from wealthier tourists, increasing their personal wealth and that of surrounding businesses. However, BanKhai has 250% more derelict land which may also indicate that it has a weaker economic situation than Thaton. The hypothesis is further substantiated as BanKhai has eight times more manufacturing shops (secondary industries) than Thaton. This relates to the Clarke-Fisher model whereby as an area develops, the secondary industry declines (reference figure 14). These findings show that Thaton is significantly more developed as proven by its greater tertiary industry employment and a correspondingly fewer manufacturing shops (supporting the hypothesis).

Table 8- Key for figure 7

General Use	Code	Types
Residential	R1	Low density
	R2	High density
Industrial	I1	Manufacturing (Small)
	I2	Manufacturing (Large)
Commercial	C1	ATM / Bank
	C2	Garages
	C3	Markets
	C4	Offices
	C5	Shops (Low order)
	C6	Shops (Med order)
	C7	Shops (High order)
	C8	Others
Public buildings	P1	Colleges / Schools
	P2	Government
	P4	Temples
	P5	Others
Entertainment	E1	Hotels / Guest houses
	E2	Restaurants / Bars
	E3	Others
Open space	O1	Car parks
	O2	Parks
	O3	Recreation
	O4	Others
Unused	U1	Derelict
	U2	Vacant
	U3	Others

Figure 14- Clarke-fisher model

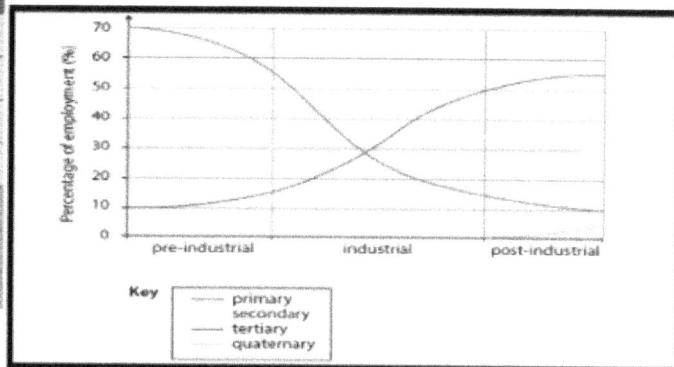

(Kokkinos, 2014)

75

Chi Squared Test

A Chi Square test can be used in order to test the relationship between two categorical variables i.e. land usage and location (Laerd Statistics, 2013).

From primary data obtained from surveys, the impact of tourism on Mae Ai's land usages can be evaluated by comparing the differences between the recorded land usages at the 25 sites. From this, it can be determined if there the tourism industry causes any significant differences to the Mae Ai industry.

Null hypothesis: There is no significant difference between the land usages in Thaton and BanKhai.

Alternate hypothesis: There is a significant difference between the land usages in Thaton and BanKhai.

Table 9- Data and formula required to obtain Chi Square values

LAND USE CLASSIFICATION	OBSERVED THATON	OBSERVED BAN KHAI	ROW TOTAL	EXPECTED (ROW TOTAL × COLUMN TOTAL / OVERALL TOTAL) THATON	EXPECTED BAN KHAI	CHI SQUARED VALUE ((O-E)²/E) THATON	CHI SQUARED VALUE BAN KHAI
R1	6	3	9	4.5	4.5	0.5	0.5
R2	6	7	13	6.5	6.5	0.04	0.04
	0	1	1	1	1	1.00	1.00
	1	6	7	3.5	3.5	1.79	1.79
C1	1	0	1	0.5	0.5	0.50	0.50
C2	1	0	1	0.5	0.5	0.50	0.50
C3	0	0	0	0	0	0.00	0.00
C4	0	0	0	0	0	0.00	0.00
C5	9	6	15	7.5	7.5	0.30	0.30
C6	2	1	3	1.5	1.5	0.17	0.17
C7	5	3	8	4	4	0.25	0.25
	2	0	2	1	1	1.00	1.00
F1	0	2	2	1	1	1.00	1.00
F2	1	2	3	1.5	1.5	0.17	0.17
F3	0	0	0	0	0	0.00	0.00
F4	1	2	3	1.5	1.5	0.17	0.17
	1	0	1	0.5	0.5	0.50	0.50
	7	3	10	5	5	0.80	0.80
	1	0	1	0.5	0.5	0.50	0.50
G1	0	0	0	0	0	0.00	0.00
G2	0	0	0	0	0	0.00	0.00
G3	0	0	0	0	0	0.00	0.00
G4	0	0	0	0	0	0.00	0.00
H1	2	5	7	3.5	3.5	0.64	0.64
	3	7	10	5	5	0.80	0.80
	1	1	2	1	1	0	0
OBSERVED COLUMN TOTAL	50	50	100			Chi squared value= 10.62	10.62

Total chi square value: 21.24

$$Degrees\ of\ freedom = (number\ of\ rows - 1) \times (number\ of\ columns - 1) = 24 \times 1 = 24$$

76

Table 10- Chi Square degrees of freedom table

Percentage Points of the Chi-Square Distribution

Degrees of Freedom	Probability of a larger value of x^2								
	0.99	0.95	0.90	0.75	0.50	0.25	0.10	0.05	0.01
1	0.000	0.004	0.016	0.102	0.455	1.32	2.71	3.84	6.63
2	0.020	0.103	0.211	0.575	1.386	2.77	4.61	5.99	9.21
3	0.115	0.352	0.584	1.212	2.366	4.11	6.25	7.81	11.34
4	0.297	0.711	1.064	1.923	3.357	5.39	7.78	9.49	13.28
5	0.554	1.145	1.610	2.675	4.351	6.63	9.24	11.07	15.09
6	0.872	1.635	2.204	3.455	5.348	7.84	10.64	12.59	16.81
7	1.239	2.167	2.833	4.255	6.346	9.04	12.02	14.07	18.48
8	1.647	2.733	3.490	5.071	7.344	10.22	13.36	15.51	20.09
9	2.088	3.325	4.168	5.899	8.343	11.39	14.68	16.92	21.67
10	2.558	3.940	4.865	6.737	9.342	12.55	15.99	18.31	23.21
11	3.053	4.575	5.578	7.584	10.341	13.70	17.28	19.68	24.72
12	3.571	5.226	6.304	8.438	11.340	14.85	18.55	21.03	26.22
13	4.107	5.892	7.042	9.299	12.340	15.98	19.81	22.36	27.69
14	4.660	6.571	7.790	10.165	13.339	17.12	21.06	23.68	29.14
15	5.229	7.261	8.547	11.037	14.339	18.25	22.31	25.00	30.58
16	5.812	7.962	9.312	11.912	15.338	19.37	23.54	26.30	32.00
17	6.408	8.672	10.085	12.792	16.338	20.49	24.77	27.59	33.41
18	7.015	9.390	10.865	13.675	17.338	21.60	25.99	28.87	34.80
19	7.633	10.117	11.651	14.562	18.338	22.72	27.20	30.14	36.19
20	8.260	10.851	12.443	15.452	19.337	23.83	28.41	31.41	37.57
22	9.542	12.338	14.041	17.240	21.337	26.04	30.81	33.92	40.29
24	10.856	13.848	15.659	(19.037)	23.337	28.24	33.20	36.42	42.98

(Laerd Statistics, 2013)

Corresponding chi-squared value on the degrees of freedom table

Uncertainty exceeds 0.75. This means the results are 25% certain that there is a significant difference between the land usages of the two locations. Therefore, the null hypothesis is accepted and there is no statistically significant difference between land usages in Thaton and BanKhai. However, at face value the type of land usages seems to differ between the two villages as shown in figure 7. Therefore, the hypothesis that tourism improves the socio-economic situation of Mae Ai (by providing different facilities such as high-order shops) is still valid.

Trends between tourist density and vehicle count:

Figure 15- Number of pedestrians per 10 minutes in Thaton and BanKhai

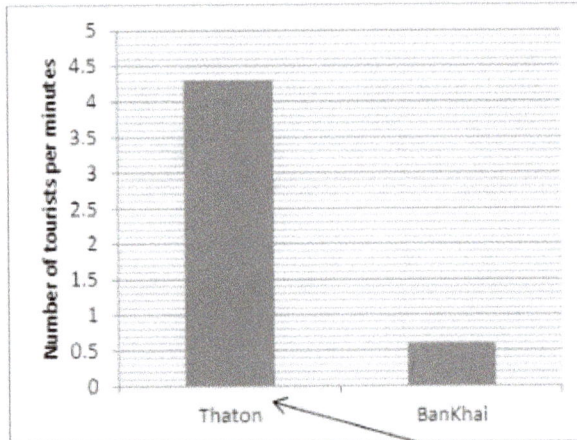

Figure 15a- Number of vehicles per 10 minutes in Thato and BanKhai

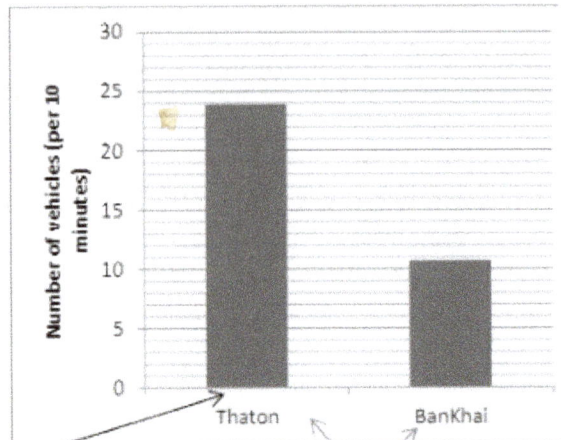

Higher human activity in Thaton than BanKhai

This difference may be explained by the Butler Model

As observed in Figures 15 and 15a, Thaton has a higher pedestrian and vehicle counts than BanKhai at 7.17 and 2.23 times repectively. This can be interpreted using the Hawkins model which investigates the factors that may have caused the perceptual carrying capacity to have been reached (Codrington, 2002). The data would suggest tourism brings about undesired congestion. However, questionnaire results (reference table 11) suggest this to be otherwise with none citing congestion as a tourism problem. This means that the carrying capacity has yet to be reached and this increased flow of people is positive.

This is because tourism revenue encourages the flow of people into Thaton (WTO, n.d.). Aditionally, the fact that more vehicles are present means the population has a greater disposable income. This data shows Thaton has entered the "success stage" on the Butler Model; thus supporting the hypothesis and bring positive socio-economic changes to Mae Ai.

Determining trends in perception of local businesses

Table 11- Questionnaire and results of tourism impacts on local businesses

Business key:

Business one

Business two

Business three

Business four

Business five

Business six

Anomaly as despite not catering to tourist business increased

The open-ended answer choices make graphing data difficult

Type of business _____

1. How long has the business been operating?

10 years

30 years

Don't understand question

3 years

15 years

4 years

*Mean = 7 years 8 months

2. Did you set up the business mainly for tourists?

Yes

No

No

Yes

Yes

Yes

3. What percentage of your business comes from tourism?

15%

10%

30%

30%

10%

60%

*Mean = 29%

4. How many people (including yourself) do you employ?

3 people

4

4

4 people

0

3

*Mean = 3 people

5. Has the amount of business increased, decreased or stayed the same over the past few years?

Increased

Decreased

Increased

Increased

Increased

Increased

5. What problems are there of relying on tourism...

Cultural differences

N/A

Low prices who spend money and may conflict with locals

Money

Language barriers

Language barriers

This results of the questionnaire shown in table 11 show that 5 in 6 shops reported an "increase" in business with 80% surveyed shops catering for tourists. This suggests that shops targeting (and thereby receiving more) tourists have higher revenue. This is in accordance with the multiplier effect as more people benefited from tourism. Similarly it changes the local perception of an "ideal" shop as when they become wealthier they seek similar goods and services as their tourist counterparts. This therefore stipulates to the further growth of tourist areas.

An anomaly was observed in Business 3 (a convenience store) that experienced an increase in economic activity despite not targeting the tourists. Nevertheless, tourism has indirectly benefitted this store as more people visited Thaton.

However, Business 2 (a noodle shop targeting locals) experienced a decrease in income recent years due to over-competition. This means traditional businesses may face competition from similar tourist orientated services (restaurants in this case). While this constitutes a minority at present, this ought to be evaluated further to see if there is a rising trend.

Despite anomalies, this data concludes that tourism does benefit the economic situation of Thaton (in Mae Ai) as all of the businesses cater to tourists and most shops overall experiencing an "increase" in business.

Determining trends in perception of individuals:

Figure 15- Questionnaire answers from Individual perception questionnaire

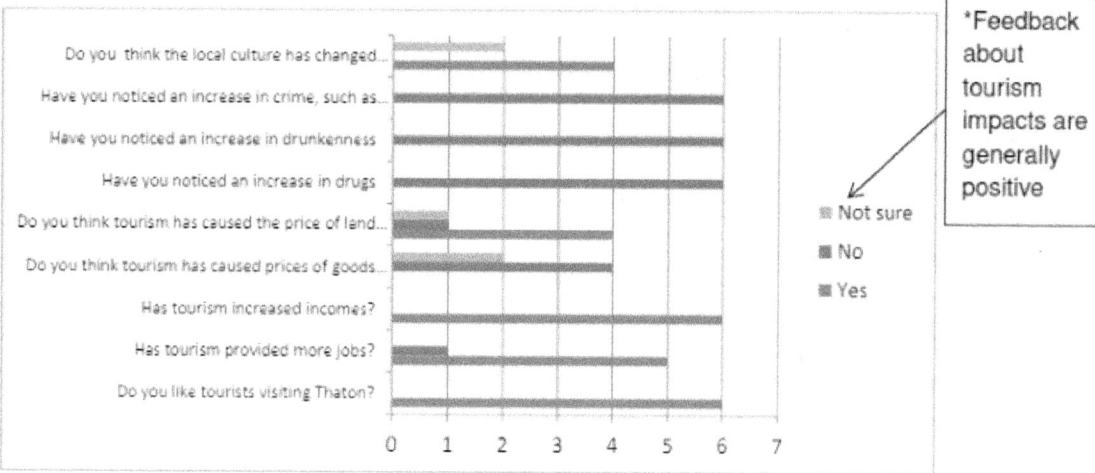

While some locals perceive tourism negatively (due to increased land prices), most enjoy the jobs, income, and tourist interactions arising from domestic tourism. Therefore, it can be concluded that the perceptual carrying capacity of Thaton has not been reached; meaning that tourism is still considered beneficial by locals (Rees, 1996).

The social effect tourism has had on culture is evident since all 7 participants admitted tourism had changed the local culture. From observations, youthful participants claim to be unsure while older participants claim that there has definitely been a cultural shift. However, as no large-scale studies have been conducted to investigate the cultural impact, one can infer from the positive answers to the last four questions (reference figure 16) that these impacts are positive.

One participant did claim that tourism increased crime. This anomaly might be due to her personal experiences of being mugged by a tourist - as she informed the surveyors. As the sample size is small (7 participants), this incident may not be anomalous, but significant. It is therefore prudent to increase sample size to confirm whether tourism brings about a rise in crime rates.

In general, however, the social impacts of tourism from this data seem favourable and the results support the hypothesis.

Trends in Quality of life:

Figure 16- Quality of life survey scores in Thaton from 20 people

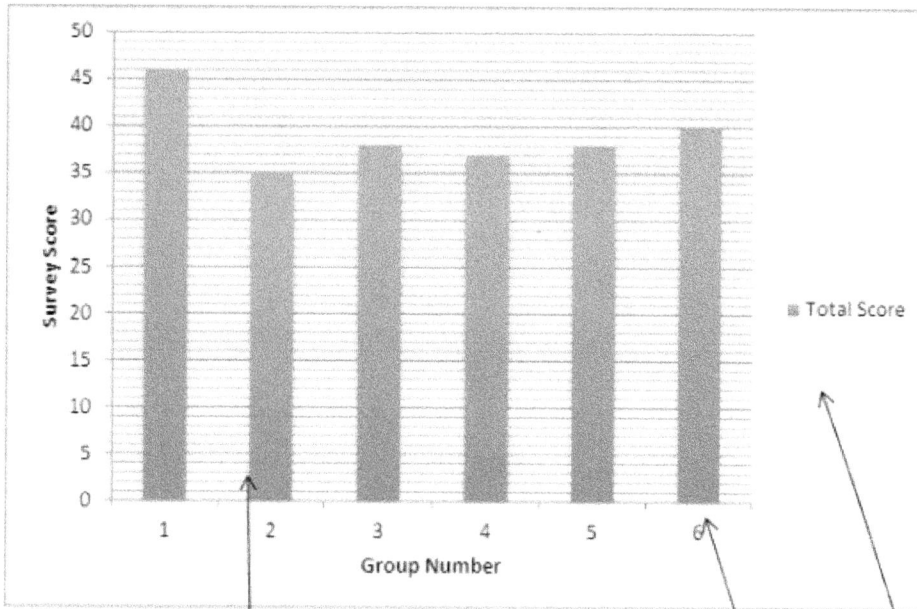

Thaton unanimously scored higher than BanKhai

Thaton mean score= 39

Maximum score refers to highest possible scoring

Figure 17- Quality of life survey scores in BanKhai from 20 people

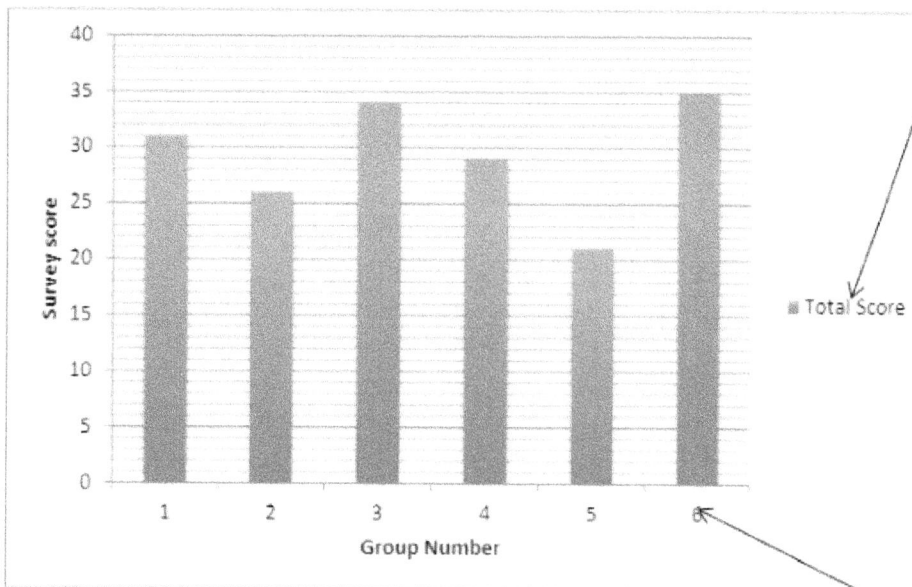

BanKhai mean score= 29

Thaton has an average score of 10 points per group higher than BanKhai with a mean score of 39 and 29 for Thaton and BanKhai respectively; this may be caused by increased tourist revenue. Thus, due to its tourism industry the environmental quality and overall quality of life improved immensely in Thaton. With its strategic location, accessible roads, river routes, religious temples and tourist attractions (reference figure 18), Thaton has embraced the tourist industry by building more hotels, guest houses and restaurants with better materials (reference figure 19). A lack of funding in BanKhai forced it to use inferior building materials which in turn led to crumbling infra-structure (reference figure 20) (Gurburz, 2016) and significantly reduced quality of life ratings (reference figure 17). This is due to the geographical situation being less favourable and since BanKhai was not in close proximity to primary tourist sites less tourism revenue is available and minimal multiplier effect induced.

Figure 18- Tourist map of Thaton

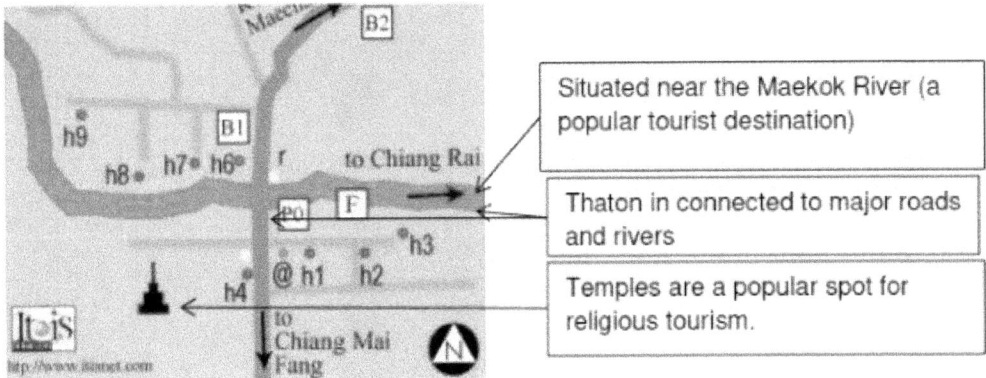

Situated near the Maekok River (a popular tourist destination)

Thaton in connected to major roads and rivers

Temples are a popular spot for religious tourism.

(ITIS, 2002)

A similar map showing the geographical situation of BanKhai is unavailable due to its relative obscurity.

84

Figure 20- Field sketch of a Thaton building

Figure 19- Photograph of a BanKhai house

Newer brick houses improved quality of life ratings

(Author's own, 2016)

(Chang, 2016)

Older rusting housing caused lower Ban Khai ratings

The illustrated evidences support the hypothesis:

Thaton has more disposable income from tourism so can afford better building materials to build better houses than BanKhai.

Trends in standard of living:

Figure 21- Thaton housing living quality mean rankings

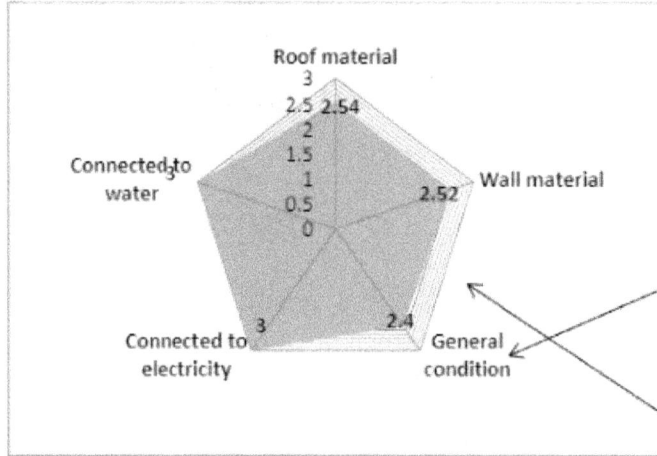

Figure 22- Ban Khai housing living quality mean ranking

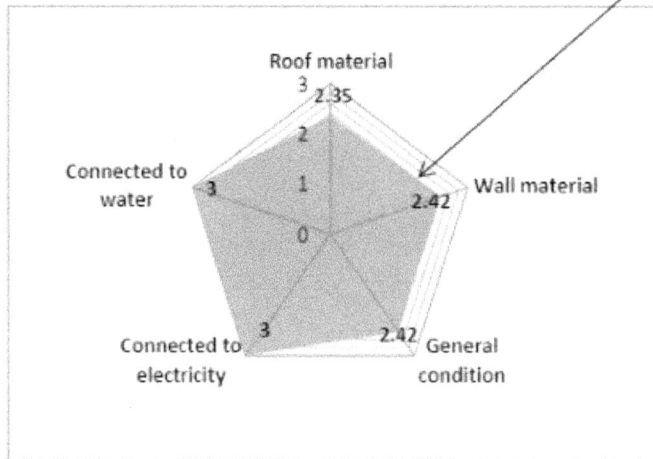

Anomaly as ranking is higher in BanKhai (unexpected result)

Calculated mean results from both villages (all categories)

In Thaton, tourist revenue into the region has raised the disposable income of individuals allowing them to construct better quality buildings. There is a tourist tax of 500 baht in Thailand (anon, 2013). This tax is re-invested into improving the living standards of locals (which makes Thaton even more appealing to visitors). This led to Thaton receiving a better ranking on categories such as Roof and Wall material due to its greater development (reference figures 21 and 22). This is an example of the classical Butler model in display, where Thaton has benefited from the increase in tourist number, thus gaining in prosperity, upgrading of infrastructure and flourishing of the village (Uysal, et al, 2012). BanKhai, unfortunately, lacks this source of revenue leading to overall lower living standards.

Anomalies include the relatively low general conditions ratings in Thaton (which is lower than BanKhai at 2.4 to 2.42). Prevalent vandalism (by tourists) (reference figure 23) potentially caused the ratings of general condition to be lower in Thaton than BanKhai. However, as this was the only standard of living category worse in Thaton, the data lends its supports to the hypothesis.

Figure 23- Evidence of graffiti (vandalism) at Thaton.

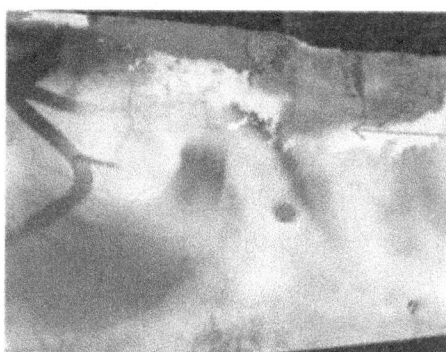

Graffiti in Thaton lowers its general condition scoring

(Chang, 2016)

Price comparison between the villages

Table 12- Prices of goods Thaton

Price comparison-	Thaton (THB)
Price of land (one rai) (Massington, 2016)	1,200,000
Price of 500ml water	15
Price of lighter	5
Price of chewing gum	5
Price of 320ml Chang beer	35
Price of biscuit	7

> Unreliable as water brands were not kept constant

> Land prices (B.Massington,2016)

Table 13- Prices of goods Ban Khai

Goods	Ban Khai (THB)
Price of land (one rai) (Massington, 2016)	400,000
Price of 500ml water	10
Price of lighter	5
Price of chewing gum	5
Price of 320ml Chang beer	33
Price of biscuit	5

Key:

Price Increase

Price Constant

Price Decrease

Tables 12 &13 show significant increases in land prices and other goods that can be justified by the multiplier-effect model. Increased tourist spending causes inflation. Increased demand leads to increases in price. Higher inflation is a significant impact of tourism (Icoz, 1991) as it can outprice locals despite what is perceived to be minor increment by tourists' standards. This survey showed that beer and water are 2 and 5 baht higher in Thaton. The most drastic difference is observed in Thaton's land price which is three times higher than nearby BanKhai due to increased land demands in this tourism hotspot. This causes increased rental prices and would subsequently lead to inflated goods prices that locals also have to pay. Tourism facilities often take priority over local needs – an occurrence that is difficult to sustain in the long term. This inflation data, unlike the others, does not support the hypothesis as inflation is a negative impact the locals will have to suffer.

Conclusion

In conclusion, the hypothesis "Tourism has a positive socio-economic impact on Mae Ai" is mostly true. The answer to what socio-economic impacts does tourism bring to rural communities in the Mae Ai region of Thailand it would be increased prosperity, quality of life, living standards but these positive impacts is countered by high inflation rates. Unfortunately, several social issues such as potential crime from tourism may have arisen, yet insufficient data was collected to address the relevance of this result.

The results support the Multiplier-effect and Butler's model which implies that tourism stimulates development, but also causes inflation and potentially congestion (reference figures 15, 15a, Tables 12 &13).

The main reason for the conclusion is that the perceptual carrying capacity has not yet been reached and the socio-economic impacts are mostly positive.

Evaluation

Table 14- Table of evaluations

Limitation	Improvement
Traffic /people counts were only conducted in the mornings/ afternoons.	Traffic and people count should occur repeatedly.
Only 25 sites were selected for sampling.	Conduct investigation on longer transect with more sites to get more representative samples.
Not enough questionnaires were distributed (only 6 businesses and 14 individuals).	Repeat more and also prevent asking similar demographic e.g. females who were often working at stores. This can be done by doing random number tables to select shops to ask and asking every other individual that passes by.
Individual subjectivity was occurred during land use survey and other qualitative observations	Conduct a pilot study (Nagle & Spencer, 1997) to standardise the rankings. This can be conducted in a different Mae Ai village.
Goods across various brands had fluctuating prices	Ensure prices are compared with same brands
Subjectivity in the quality of land use survey.	Photographic evidence allows more to evaluate and score these locations (minimising subjectivity)
Economic impact is judged by merely three factors	Other qualitative factors should be included in the survey and more factors e.g. environmental impacts.
Many struggled to comprehend the questionnaire due to the poor translations	Employing a translator to aurally translate would widen the questionnaire scopes

Alternative Investigations

This investigation has demonstrated that the social economic impacts of tourism are generally positive for Mae Ai given that the perceptual carrying capacity has yet to be reached. It would be highly interesting to repeat this investigation in a few years when the tourist counts is likely to have increased and investigate whether or not impacts were still similar or if the perceptual carrying capacity had been exceeded.

Alternatively, one could investigate the environmental impacts tourism brings to the Mae Ai community to investigate whether or not the environmental carrying capacity of Thaton has been reached.

An alternative question could be "What are the environmental impacts of tourism on the region of Mae Ai?" and information such as soil degradation at tourist hotspots e.g. around temples/ rivers could be collected and evaluated.

BIBLIOGRAPHY + APPENDICES OMITTED

3. EXAMPLE THREE (22/25)

Title: 'What is the pattern of land use in Mirów, Warsaw

and what are potential factors that affect it?

Author: Anonymous

Session: May 2022

Level: SL

Examiner's summary

Criterion A [3/3]:

The student has developed a narrowly focused geographical fieldwork question and clearly stated the location of the study. The presented geographical theory, including background information and concepts, is relevant to the exploration. The fieldwork question is explored through a collection of primary and secondary data, and the predictions/hypotheses are well-developed and explained. However, the student did not provide a justification for the choice of fieldwork location. The geographical context is provided, and the topic of the study is linked to the syllabus. The student included one or more locational maps with a title, labels, scale, cardinal directions, and a key.

Criterion B [2/3]:

The student provided a detailed description of the methods chosen for both primary and secondary data collection, including the most suitable techniques for collecting quality data. Additionally, the student included all necessary figures such as pictures, maps, or sample worksheets to explain the methods of investigation. However, the student did not provide a justification for the chosen methods of data collection. The date, time, and location where the data collection was conducted were also stated.

Criterion C [6/6]:

The student collected and presented enough data relevant to the fieldwork question. They included relevant maps, graphs, diagrams, and annotated photos to represent the collected spatial data, which were well-labelled and personalized. The sample size used for the analysis was sufficient and the student used a variety of techniques to present the collected data, including visual tools and written analysis.

Criterion D [8/8]:

The student provided a thorough and detailed discussion about the collected data, correctly interpreted it to identify different land uses and recognized significant trends and patterns. They also identified important outliers/anomalies and suggested their potential source. Descriptive statistics were appropriately used to assess the type and proportions of different land uses in the investigated zones. The student referred to the geographical and historical context of the investigated area and remained focused on the 3 hypotheses and the main fieldwork question.

Criterion E [2/2]:

The student has provided a clear and well-supported conclusion that answers the fieldwork question. They have referred to their original predictions/hypotheses and compared them with the findings. The student has also accurately summarized the results of the fieldwork investigation. Furthermore, the conclusion reached by the student is supported by the data collected and its analysis, as evidenced by the analysis of the land uses in Mirów.

Criterion F [1/3]:

The student demonstrated a clear understanding of the strengths and weaknesses of the fieldwork methodology used and recognized potential factors that could have affected the reliability of the collected data. However, the student did not suggest any specific improvements to enhance the methodology, despite the weaknesses implying them. Additionally, the student did not provide any explanation or justification for any suggested improvements.

Other requirements [0/0]:

The student has successfully met the requirements for the criterion as they have not exceeded the word limit of 2500 words, numbered the pages, and included references to all external sources of information.

1. Fieldwork question and geographical background

1.1. Fieldwork question

What is the pattern of land use in Mirów, Warsaw and what are potential factors that affect it?

1.2. Geographical background

Mirów is a part of the Wola district in Warsaw, Poland, that is situated close to Warsaw CBD[1]. Before the Second World War, it was characterised by extremely densely placed tenements, most of which were razed to the ground after the Warsaw Uprising. Since 1945 in this area many soc-realist residential units were built alongside large developing industrial sites. For the last two decades, the predominantly industrial and soc-realist residential character of this city zone began changing into a more CBD-like neighbourhood, with more space devoted to offices and retail, with a dozen modern construction investments being finished in the last couple of years.

1.3. Hypotheses

1) Areas closer to CBD will exhibit a greater proportion of service sites and fewer residential units.

   According to the bid rent theory, the price of land decreases with the increase in distance from the CBD[2]. Hence, the closer to the city centre, the more space is devoted to offices and retail use. It can be predicted that there will be relatively more service buildings than residential, public, or industrial units in the areas closer to Warsaw CBD due to theoretically higher prices of land. The proportions are more likely to change in the zones further from the city centre, where residential land use has bigger profitability than offices and retail.

2) Historically industrial areas are more likely to have a big share of services units.

   During the Polish People's Republic period Mirów was a part of a highly industrialised area unofficially called the West Industrial District[3]. The change of the political system in Poland in 1989 began an era of rapid socio-economic changes which significantly impacted the structure of Polish cities, Warsaw included. One of the most prominent features of the post-socialist urban transformation process was urban de-industrialisation followed by urban renewal with the development of the tertiary sector[4]. Therefore, it can be expected that zones, which were historically the most industrialised, are now mostly sites of service buildings or modern residential investments.

3) Buildings located directly by larger streets will more likely carry a non-residential function. Large, busy transportation routes are sources of noise pollution. As it is especially undesirable for residential areas, it can be predicted that buildings in direct proximity to larger streets, i.e. dual carriageways, will be predominantly services, including retail and offices, sites of industry, or public institutions.

1.4. Link to syllabus

Option G - Urban Environments

- Unit 1 – "Factors affecting the pattern of urban economic activities (retail, commercial, industrial), including physical factors, land values, proximity to a central business district (CBD) and planning."

- Unit 2 – "Urbanisation, natural increase and centripetal population movements, including rural-urban migration in industrialising cities, and inner-city gentrification in post-industrial cities."

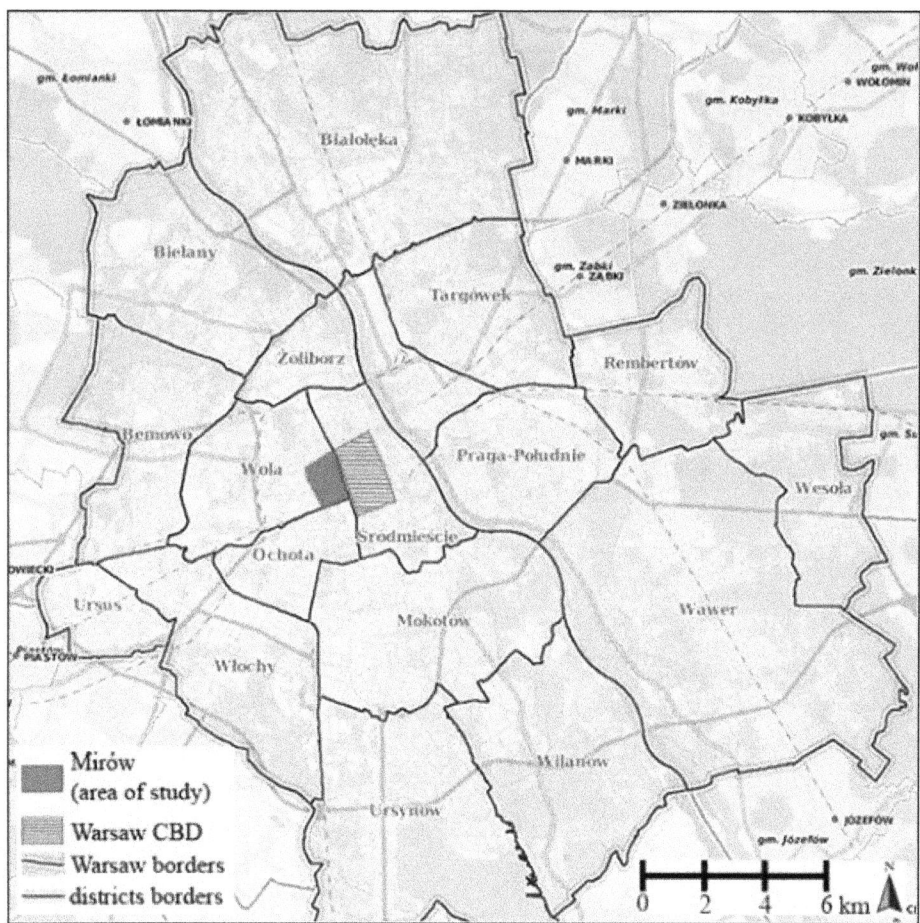

Map 1. Warsaw with labelled area of study

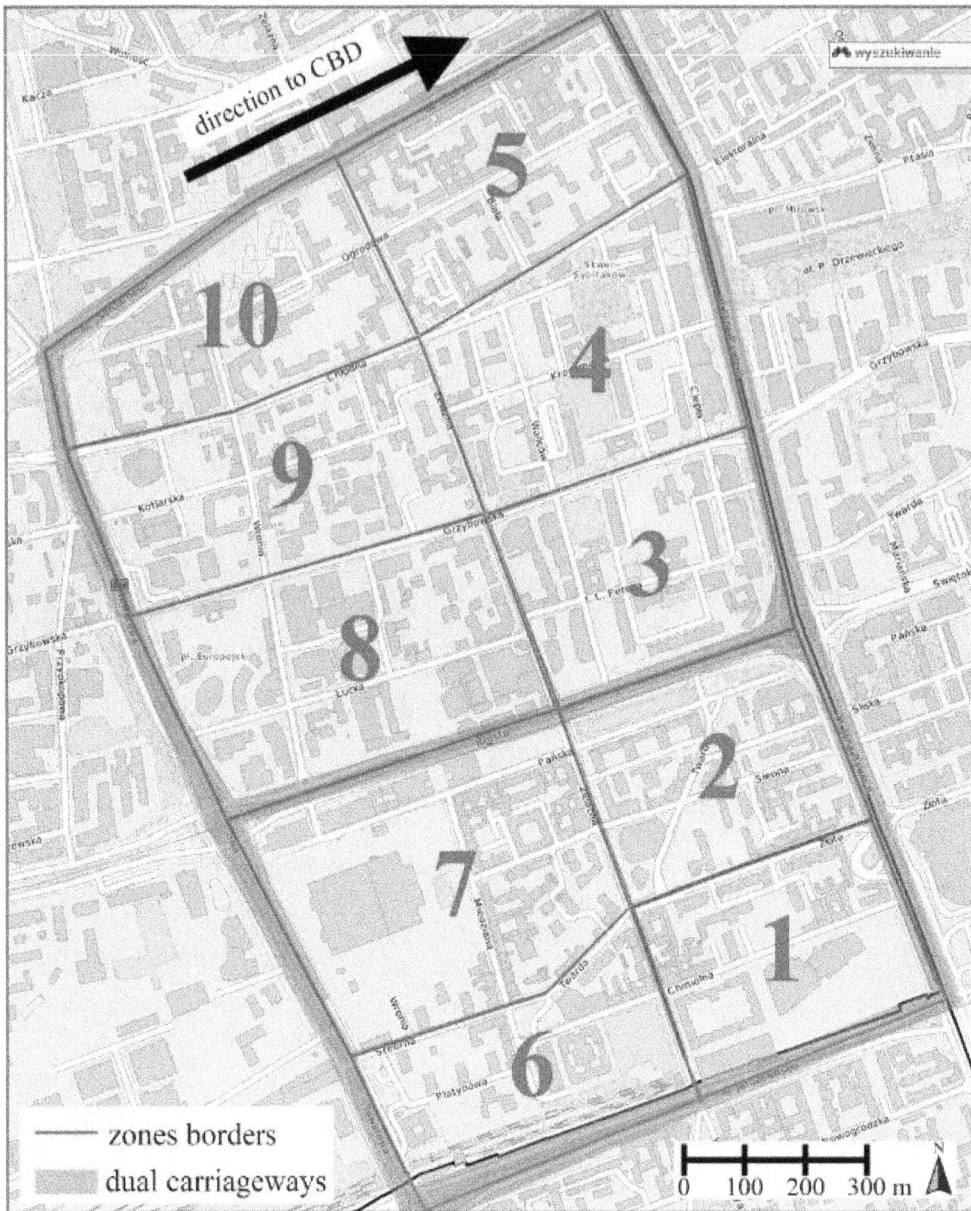

Map 2. Division of Mirów into zones

2. Methodology

The research was conducted on June 9th and 10th, 2021 (9 am - 2 pm). The area of study was set as the entire district of Mirów, which was divided into 10 zones of approximately similar area (mean: 15.21 ha; SD: 2.62 ha) (Map 2.). Zones 1-5 are directly next to a Warsaw CBD area, whereas zones 6-10 are located further away. Each zone was examined in terms of urban and architectural development, including several factors.

2.1. Land use survey

Each building was carefully observed and assigned with a specific urban function that it performs. Dwellings were divided into 5 major groups, defined as:

- Sites of services - structures located on a parcel of commercial real estate intended to generate a profit, including retail sites, offices, hotels and blocks of rental apartments, restaurants.

- Residential buildings - multifamily housing, that is tenements, apartment blocks

- Public institutions - buildings devoted to civic/community uses, including schools, courts, museums, administrative structures, churches.

- Industrial sites - sites of production, storage, or electricity substations.

- Abandoned buildings.

Buildings showing vertical zoning were assigned to the category which occupied a greater share of the levels so that each building was ascribed with one urban function.

Developments under construction were not included in the study unless their state enabled determining their function.

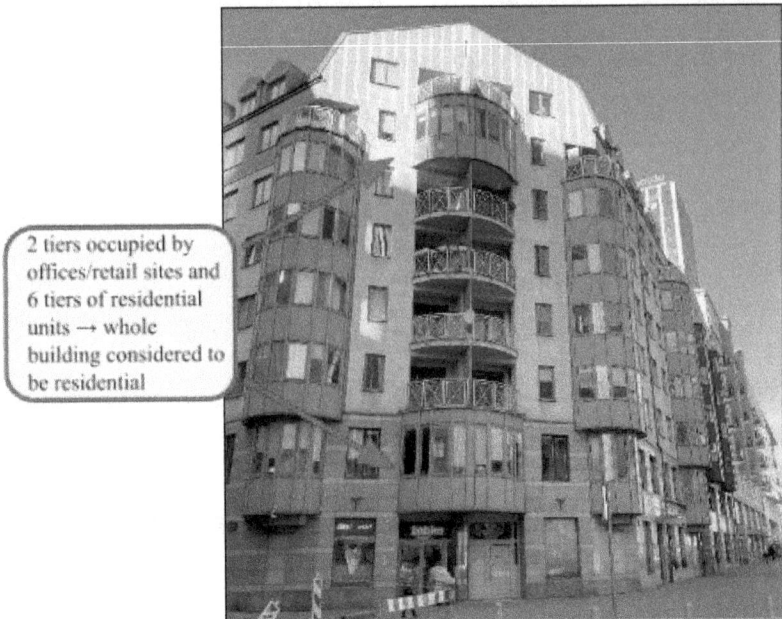

Picture 1. Example of a classification of a building with vertical zoning

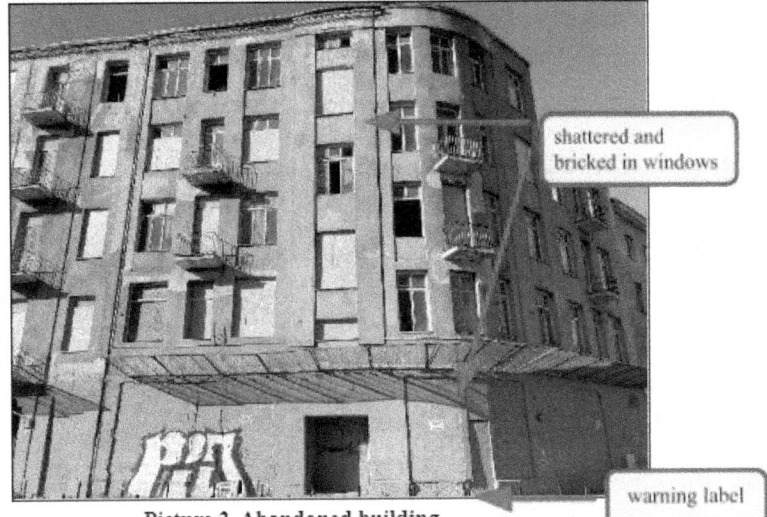

Picture 2. Abandoned building

2.1. Residential buildings classification

Residential buildings were further divided according to the historical period and style they were built in. The classification was based on the exterior characteristics of the buildings. Three categories of residential units were as follows:

- Pre-war tenements - both renovated and in the original conditions (Picture 3.)

- Socialist realist housings - Plattenbau buildings in a variety of heights, organised mostly in larger estates (Picture 4.)

- Modern apartment buildings - investments developed post-Polska Ludowa period, including those built in the late 20th century, as well as recently finished construction projects (Picture 5.)

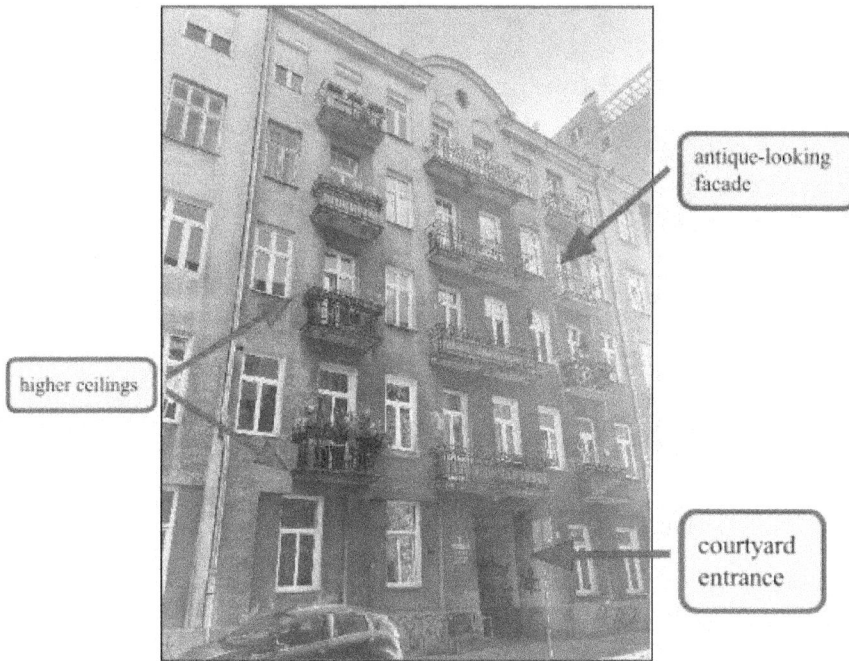

antique-looking facade

higher ceilings

courtyard entrance

Picture 3. Pre-war tenement

constructed of large, prefabricated concrete slabs

stark look of the facade

repetitive layout of windows

Picture 4. Soc-realist block

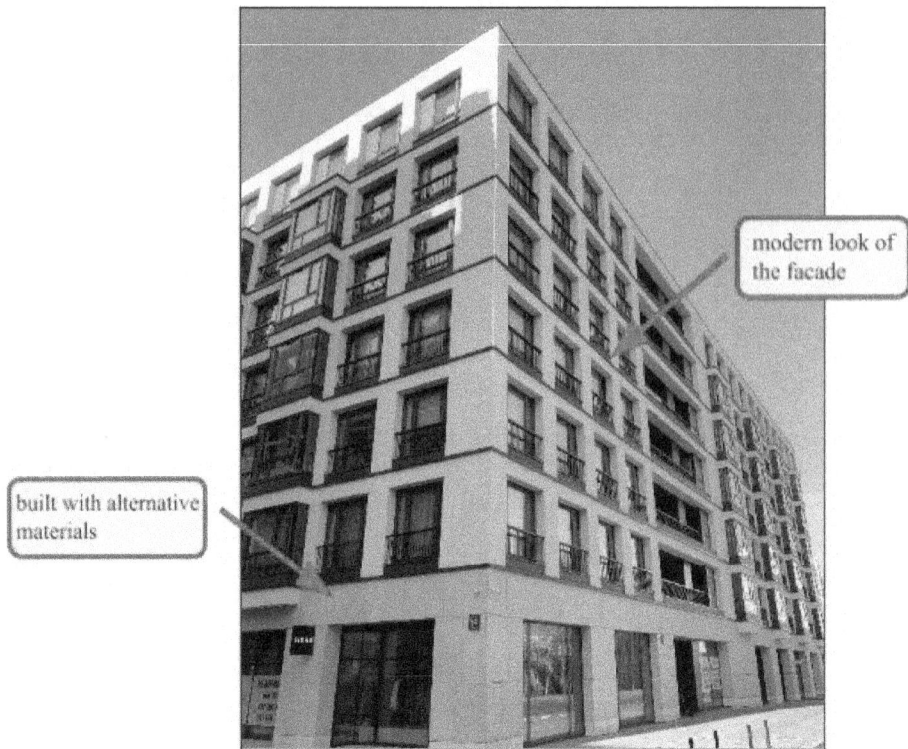

Picture 5. Modern residential building

<u>2.3. Secondary data</u>

Historical data was obtained from Warsaw City Hall website[5] and in-depth research on Mirów district organisation in the past. Locations of industrial sites functioning in the latter part of the 20th century were established.

3. Results and analysis

3.1. Historical industrial areas

Map 3. Historical sites of industry in Mirów

Historical data on the location of industrial sites in Mirów was established and depicted on the map (Map 3.). It would provide a basis to test Hypothesis 2. One can notice that zones with the biggest share of the area occupied by industrial properties are zones 8 and 9. Three other sectors that had some production sites were zones 3, 7, and 10.

3.2. Distribution of buildings of different urban functions

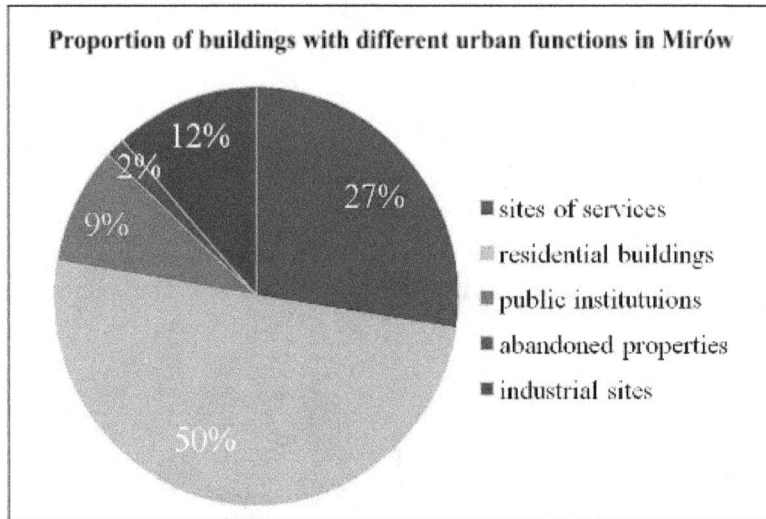

Graph 1. Proportion of buildings with different urban functions in the whole Mirów district

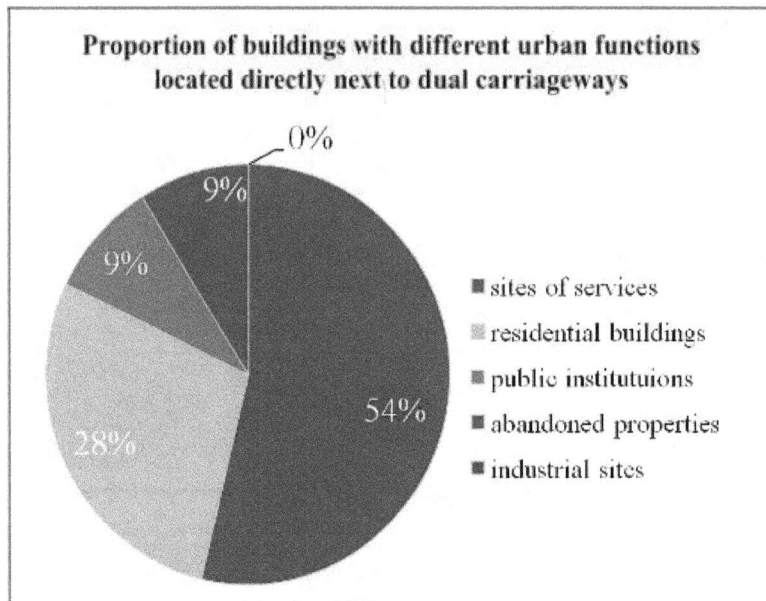

Graph 2. Proportion of buildings with different urban functions situated directly next to dual carriageways

104

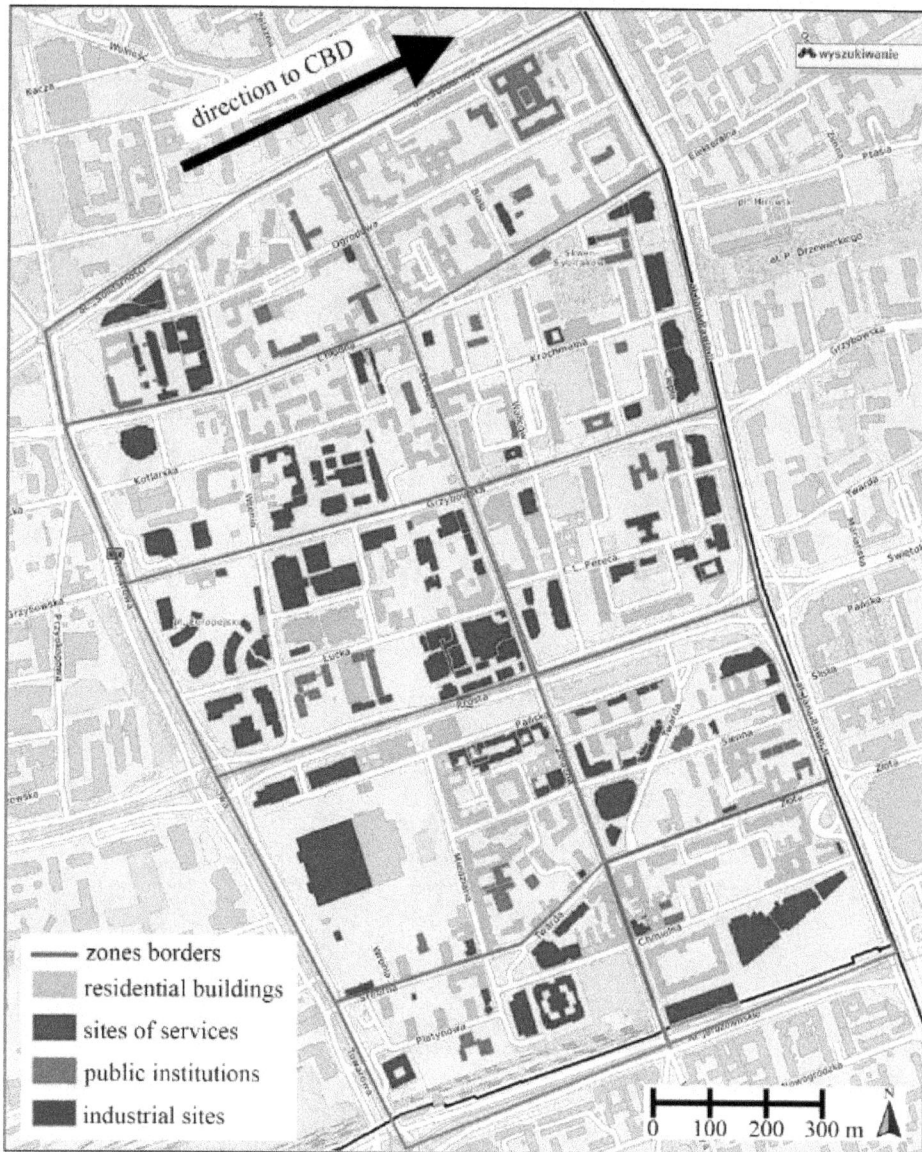

Map 4. Mirów; urban function of buildings

legend:
- zones borders
- residential buildings
- sites of services
- public institutions
- industrial sites

direction to CBD

wyszukiwanie

0   100   200   300 m

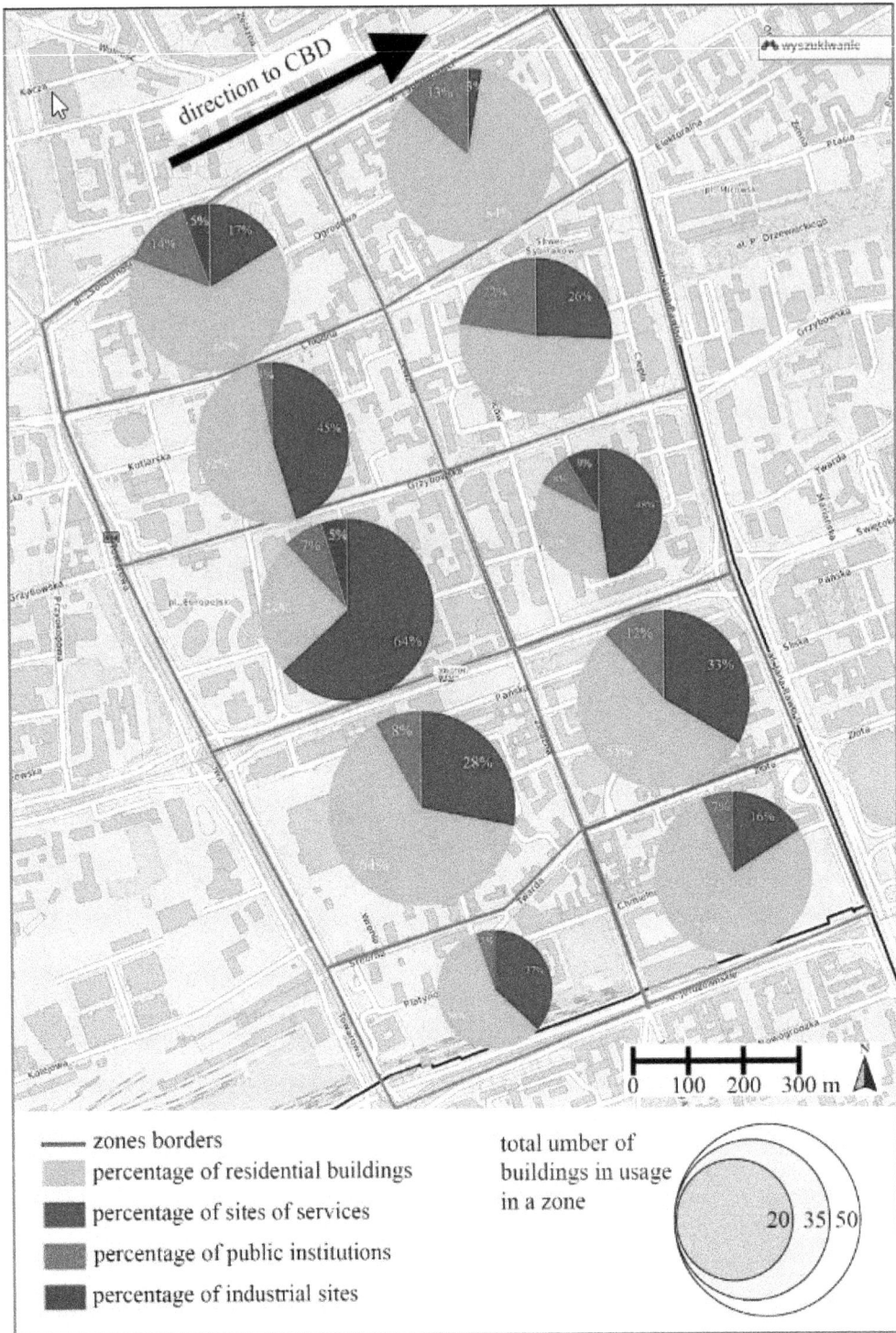

direction to CBD

wyszukiwanie

0   100   200   300 m

zones borders
percentage of residential buildings
percentage of sites of services
percentage of public institutions
percentage of industrial sites

total umber of
buildings in usage
in a zone

20 35 50

Map 5. Proportion of buildings of different urban function in each zone

Information on buildings' functions was depicted on two maps. One plan of Mirów illustrates a detailed arrangement of buildings and their function (Map 4.), whereas the other one shows what are the proportions of urban functions in each of 10 zones (Map 5.), calculated from raw data (Appendix 1.). Based on these pictures, several observations could be made.

Firstly, there is a pattern of commercial buildings arrangement, most of which are situated by main roads, i.e. Jana Pawła II, Aleje Jerozolimskie, Prosta, and Towarowa. This pattern is supported by the increased share of commercial buildings, when only buildings located directly next to dual carriageways are considered 54% (Graph 2.), compared to their percentage in the whole Mirów district - 27% (Graph 1.). Small services sites are also located further from these streets and their distribution does not have any apparent pattern. However, there is a larger aggregation of commercial buildings in Zones 8 and 9 next to Grzybowska street. Moreover, these two sectors have proportionally the most commercial sites, which add up to over 60% of buildings in Zone 8 and almost 50% in Zone 9. Additionally, Zone 3 has a significant share of services (48%). Comparing maps of past industry (Map 3.) and present urban functions (Map 4.), one can conclude that the majority of areas of dense commercial development are situated in the sites of historically industrial plants.

Furthermore, pairs of zones (1-6, 2-7, 3-8, 4-9, 5-10) in which one is closer to the CBD than the other can be compared. In such a situation, it can be recognised that zones adjacent to the CBD area (1-5) have relatively fewer commercial sites than sectors located further away. This trend is correct for all pairs but one, as Zone 2 has a slightly greater proportion of services (31.0%) than Zone 7 does (27.5%).

Public spaces are uniformly distributed throughout the whole Mirów district. Although some differences in their share between zones are present, they are mostly due to differences in the number of buildings for other functions. It is reasonable, as many of them are schools, medical centres, and administrative stations, which are usually evenly allocated, to ensure that all citizens live in relative proximity to them, as they can be considered low-to-middle order services.

Lastly, there is no industrial activity in the majority of zones. Only zones 3, 8, and 10 have a couple of buildings of this function, almost all of which are electrical substations, the only exception being the site of pharmaceutical production located in Zone 10. This is the only factory that is still active in the entire district.

3.3. Distribution of different types of residential buildings

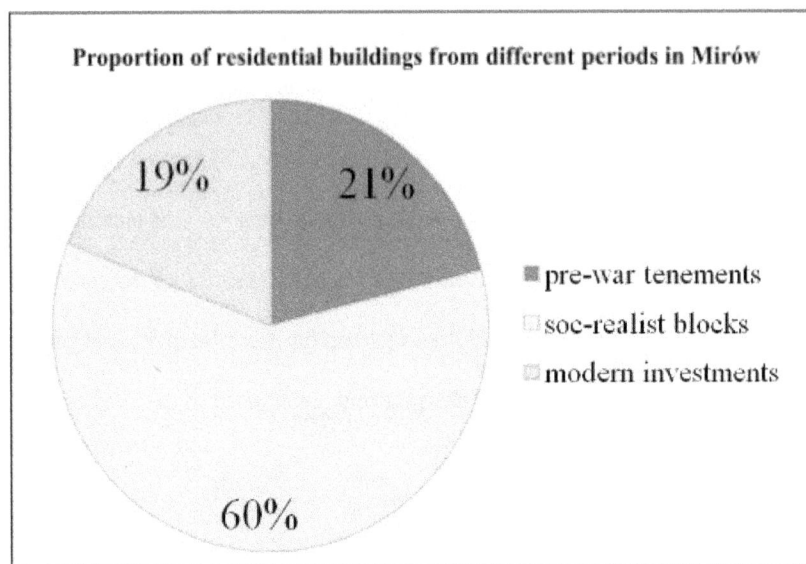

Proportion of residential buildings from different periods in Mirów

19%   21%

60%

- pre-war tenements
- soc-realist blocks
- modern investments

Graph 3. Proportion of different types of residential buildings in the whole Mirów district

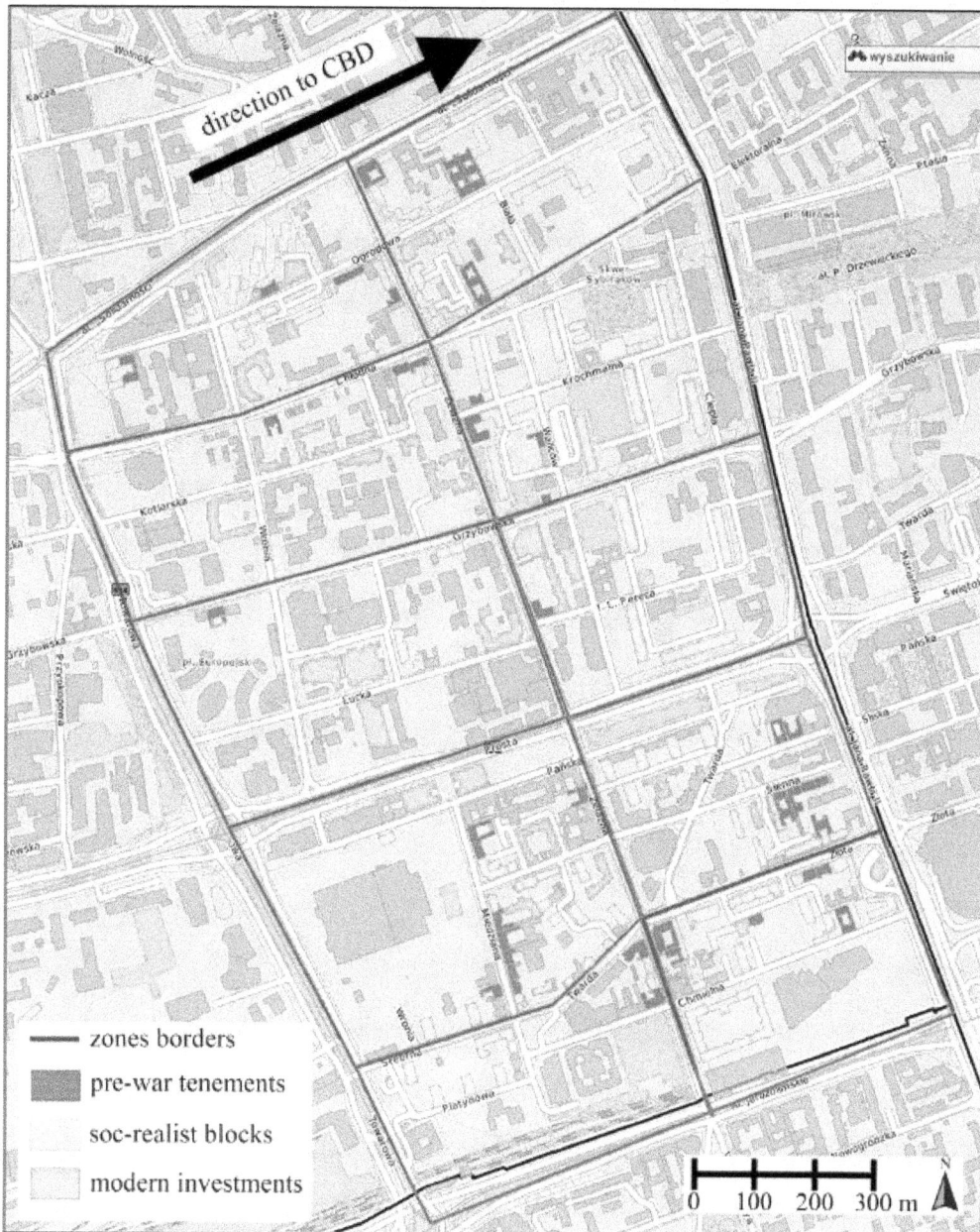

Map 6. Mirów; residential buildings from different periods

zones borders
pre-war tenements
soc-realist blocks
modern investments

direction to CBD

0   100   200   300 m

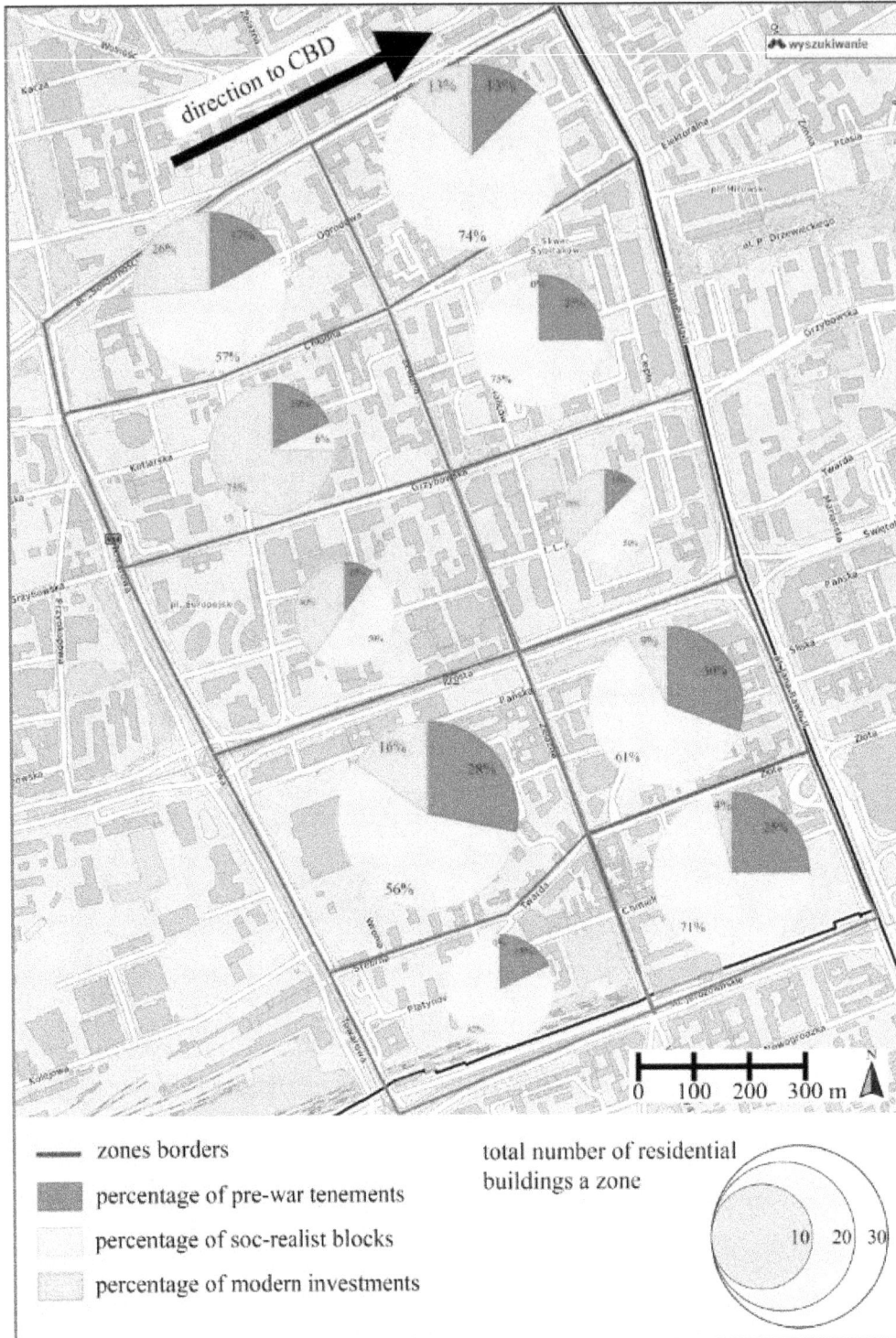

Map 7. Proportion of residential buildings from different periods in each zone

Two other plans of Mirów focused on the housing conditions, were set up. As previously qualitative (Map 6..) and quantitative (Map 7.) data was presented (raw data → Appendix 2.).

Residential units in Mirów are rarely situated directly by the main roads, as it was determined that only 28.4% of buildings situated directly next to dual carriageways were of that function, compared to an overall share of all buildings being residential - 57%. Moreover, they are often arranged in subdivisions of similar buildings. Generally, the most popular residential buildings are soc-realist blocks built in the Polska Ludowa period - with 60% of all residential units originating from these times. They constitute the majority of houses in all but one of the zones, the exception being Zone 9, with only one building of that standard and 75% of residential buildings constructed only in more recent years. The other two sectors in which panel buildings have the lowest, but still great share (50%) are Zone 3 and Zone 8. These results correspond to the above-analysed data, as it can be noticed that Zones 3, 8, and 9 have the lowest share of residential properties, out of which relatively few are soc-realist blocks, and proportionally many are modern residential investments. In contrast to this, zones 1, 4, and 6 have little to no share of present-day buildings.

Pre-World War II tenements are present in all segments and are mostly randomly located between younger structures. This type of housing is the most numerous in zones 2 and 7. Localisation of tenements is unplanned as it is the result of the ravage of Warsaw during the Second World War. Very few buildings prevailed, and all of them are heritage-listed, hence partially controlled by local or national administration[6]. This supervision affects the pattern of city zoning in Mirów, as any changes to these tenements have to be first approved by authorities.

3.4. Distribution of abandoned buildings

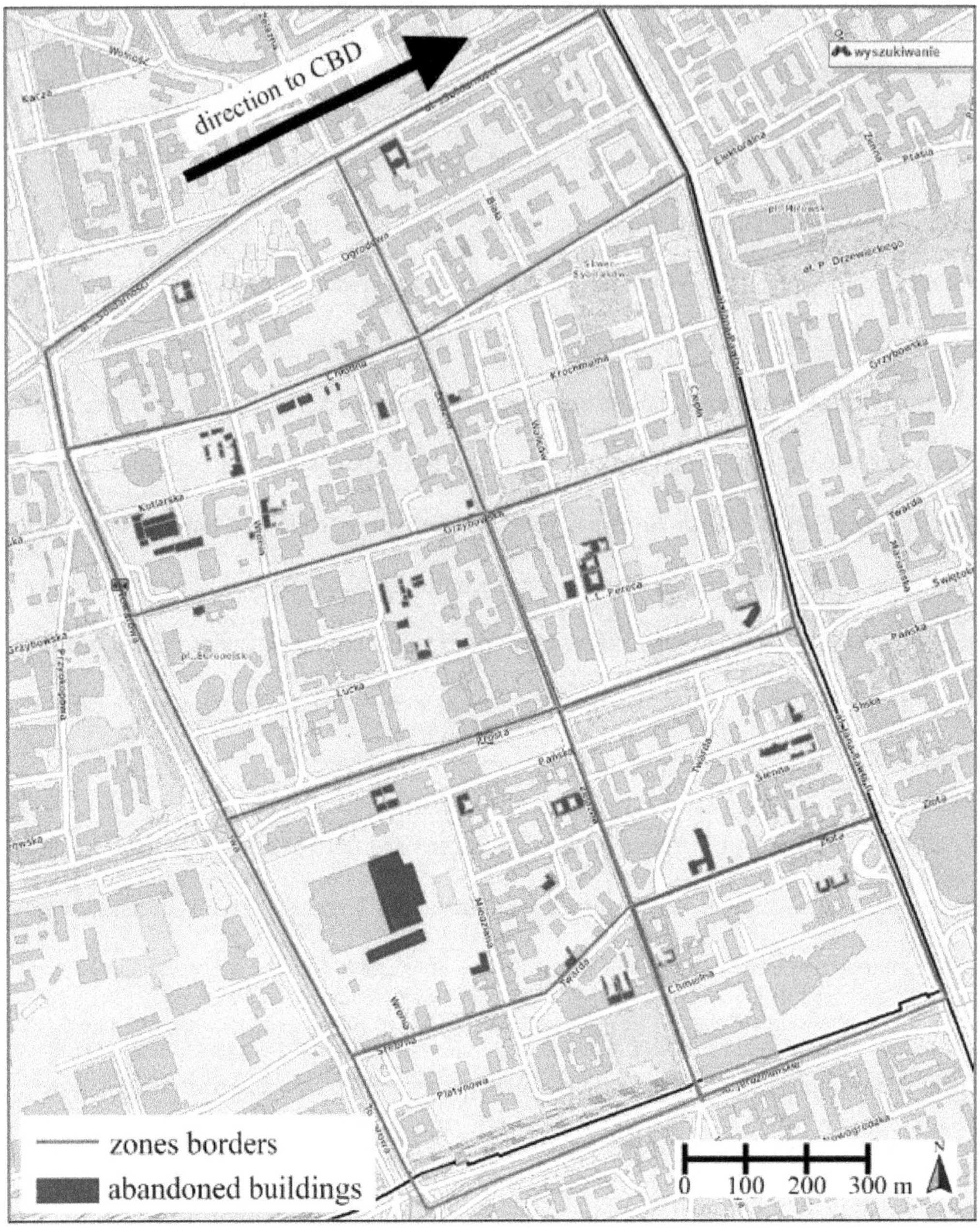

Map 8. Mirów; abandoned buildings

Although Mirów has some modern neighbourhoods and well-organised spaces, abandoned buildings are a common sight in this district. These are mostly old deserted tenements (Picture 2.) and some relatively recently closed industrial sites.

112

The biggest number of abandoned buildings is in zones 7, 8, and 9, which have 8, 7, and 11 abandoned properties correspondingly, followed by zones 2 and 3. These results can be interestingly compared with the results from previous parts of the study. What could be noticed is that a few sectors, precisely zones 3, 8, and 9 have relatively large proportions of commercial, modern residential, and abandoned buildings. It might be considered that, as these areas began to develop only recently, there was much space for new commercial and residential investments, but as they are still in a state of rapid development, there are still many dilapidated buildings.

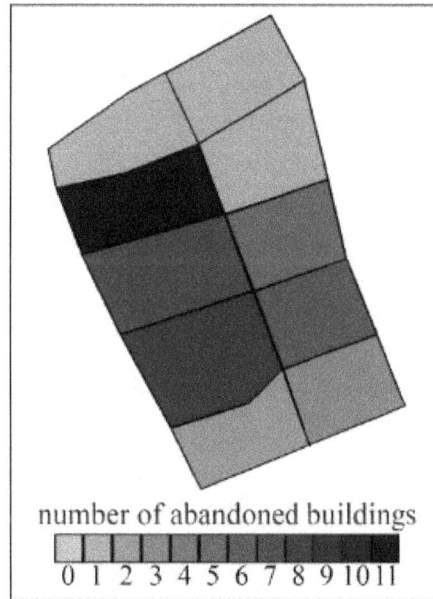

Fig. 14. Simplified plan of Mirów;
number of abandoned buildings in each zone

3.5. Assessment of hypotheses

Hypothesis 1 is fully rejected. Data did not prove that areas closer to the CBD have greater shares of commercial sites. On the contrary, the information presented in Figure 10. suggests the opposite correlation. Although the results do not fit into the bid rent curve theory, it can be justified by the fact that the differences in the distance from the city centre between zones are very small, hence almost neglectable. This model would probably be more appropriate if a larger area were studied.

113

Examining smaller fields, such as Mirów, other factors have a greater influence on the city zoning of the area.

Hypothesis 2 was to a large degree confirmed by the data. Zones that were strongly industrialised during the Polska Ludowa Period (Zones 8 and 9) have the highest percentages of both commercial and modern residential units. These results suggest that, after the deindustrialisation process that happened in the 1990', these parts of the Mirów district experienced urban redevelopment. The trend is not fitting in zone 6, although it was a heavily industrialised area back in this time. It could be explained by the relatively recent closure (2010) of the Graphical Establishment "Dom Słowa Polskiego" that used to be located there[7]. Additionally, although there used to be a large industrial production in this sector, its East-most part was predominantly residential with a great share of pre-war tenements. Thus, there was not much free space for modern developments. The hypothesis is further supported, as sectors that developed predominantly residentially during that Polish People's Republic period (Zones 1, 2, 4, 5), have much lower shares of commercial sites.

Hypothesis 3 was confirmed. Buildings in Mirów located in direct proximity to large streets, defined as dual carriageways, have predominantly commercial functions. Moreover, the proportion of residential units next to busy roads is lower than the overall share of this type of buildings in the Mirów district.

4. Conclusion

The fieldwork investigation showed that the pattern of land use in Mirów, Warsaw is not very ordered, and some of the buildings of different urban functions are located next to one another without any strong justification. Nevertheless, some factors have been shown to influence the spatial development of this neighbourhood, whereas others were found to be insignificant.

The main factor that influences the type of buildings in different zones of Mirów is the historical pattern of land use. It was noticed that historically industrial areas have been undergoing a process of urban redevelopment, which resulted in the development of predominantly commercial and

114

modern residential units. However, it was also shown that those zones still have the highest numbers of abandoned buildings, which are mostly residues of past brownfield sites. Oppositely, sub-districts of Mirów that developed residentially in the Polska Ludowa Period did not undergo major changes in spatial organisation since then, the sign of that being a big proportion of residential buildings from these years in those areas.

Another factor that was proven to be correlated with the pattern of land use in Mirów is the proximity of large streets to the sites. As predicted, buildings in direct adjacency to dual carriageways were on average more likely to carry a commercial than residential function, in contrast to the overall proportion of different urban functions in Mirów.

A factor that was shown to not affect the distribution of buildings with different functions was the distance from the CBD. It has been concluded that the differences in this variable between zones were too small to have an impact. Consequently, it should be recognised that the bid rent theory applies to studies covering larger areas.

5. Evaluation

The methodology of the study was detailed and allowed for analysis from multiple perspectives. It led to results, which could easily test out the hypotheses. The biggest asset of the investigation procedure was the amount of data collected. All buildings in the area were studied, which greatly eliminated the over-simplification of the outcome. The graphical representation of data was well prepared and made it easier to interpret.

Although the study was based on valid theories and was conveyed with great respect to detail, there were some ways that the methods could affect the results.

Despite careful observations being made, some buildings could be incorrectly classified into an urban function category. A similar error could happen during the categorisation of certain residential buildings, as houses built at the turn of two historical periods could have characteristics of both styles. Although some standardisation of observations was introduced, the human factor could not have been eliminated. This standardisation included an assumption to overlook the vertical zoning of the buildings if such existed, which influenced the results, as some buildings had more than one urban function, but were classified only to one category.

Collected data did not allow for any statistical tests. Therefore, all trends in land use patterns that were observed were based on visual analysis of maps, and a simple comparison of numerical values on pie charts. Consequently, the outcomes of the study are not strongly supported by statistical measures.

Although many features of buildings were considered, two important characteristics, precisely height and square footage were not the subjects of the study. As the size of buildings was not taken into account it caused a flaw in the study, as a small tenement and a large development were both counted as one residential unit. Similarly, ignoring the height of buildings, one-storey structures were not distinguished from skyscrapers. Both these factors were impossible to quantify without the help of external sources. Nevertheless, their inclusion of them in the study could imply that different trends in Mirów development would arise.

BIBLIOGRAPHY + APPENDICES OMITTED

4. EXAMPLE FOUR (24/25)

Title: Does the urban heat island effect occur in Bydgoszcz?

Author: A. Khangil

Session: May 2022

Level: HL

Examiner's summary

Criterion A [2/3]:

The student has successfully formulated a narrowly focused geographical fieldwork question that is explored through a collection of primary and secondary data, including measurements and their analysis. They have also provided a clear justification for the choice of location and predictions/hypotheses regarding the results of the fieldwork that are well-developed and explained. The student has included all necessary background information and concepts related to the exploration of the fieldwork question, including a locational map, although the provided map lacks a cardinal direction and key. Additionally, the student has identified and explained how the topic of the conducted study is linked to the areas of the geography syllabus.

Criterion B [3/3]:

The student demonstrated a precise understanding of the methods used for both primary and secondary data collection and provided a clear justification for their selection. They also chose the most appropriate techniques to collect quality data for subsequent analysis and included all necessary figures such as pictures, maps, or sample worksheets to explain the methods of the investigation. Additionally, the student stated the date, time, and location where the data collection was conducted.

Criterion C [6/6]:

The student has successfully collected and presented relevant data for the fieldwork question, including spacial data represented through maps, graphs, diagrams, and annotated photos. All figures used include appropriate labels, titles, cardinal directions, and a scale. The sample size of 17 sites on 12 occasions is sufficient for a detailed analysis, and the maps used are well-annotated and personalized. The student has effectively used a variety of techniques, such as graphs, diagrams, and tables, to present the data in a clear manner.

Criterion D [8/8]:

The student demonstrated a thorough understanding and interpretation of the collected data, identifying significant trends and patterns, and outliers/anomalies while suggesting their potential sources. The student appropriately selected descriptive techniques and applied statistical tests to the data collected and the fieldwork question. Additionally, the student referred to geographical context and theory, and provided a relevant analysis to the posed fieldwork question and assumed predictions/hypotheses.

Criterion E [2/2]:

The student's conclusion is well-supported by the data they collected and analyzed. They clearly answered the fieldwork question and referred to their original predictions/hypotheses, comparing them with the findings. Additionally, the student provided a summary of the results of the investigation.

Criterion F [3/3]:

The student demonstrated a strong understanding of the strengths and weaknesses of the fieldwork methodology used. They correctly identified potential factors that could have affected the reliability of the collected data and suggested specific and justifiable

Other requirements [0/0]:

The student did not adhere to the word limit of 2500 words, and also failed to number the pages. However, they have included references to all external sources of information.

INDRODUCTION

Urban Heat Island[1] usually occurs in big cities where a big part of the land is covered by buildings and pavements and other structures that absorb and retain heat. Urban heat island will lead to higher temperatures in the city centers rather than the suburbs and forested areas. Also in urban centers, trees and vegetation are cleared, so it is harder for the CO_2 to get absorbed.

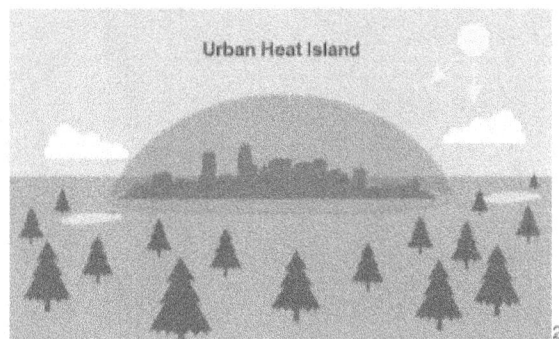

For the sake of this topic, I thought that a good idea would be to measure the temperature of different areas in Bydgoszcz. Bydgoszcz is the eighth biggest city in Poland and it has a population of 354 thousand people. During this assessment, I will examine whether the Urban Heat Island affects this city's temperature or not. This idea was created during one of our Geography lessons as climate change is the core of the Geography IB program. On the map below we can see Bydgoszcz, which is located in the northern part of Poland:

---

[1] National Geographic Society. "Urban Heat Island." National Geographic Society, 9 Oct. 2012, https://www.nationalgeographic.org/encyclopedia/urban-heat-island/.

[2] Cassette, Ian. "Roanoke to Take Part in 'Groundbreaking' Climate Research Project on Urban Heat This Summer." Https://Www.wdbj7.Com, https://www.wdbj7.com/content/news/Roanoke-to-take-part-in-groundbreaking-climate-research-project-on-urban-heat-this-summer-571161471.html.

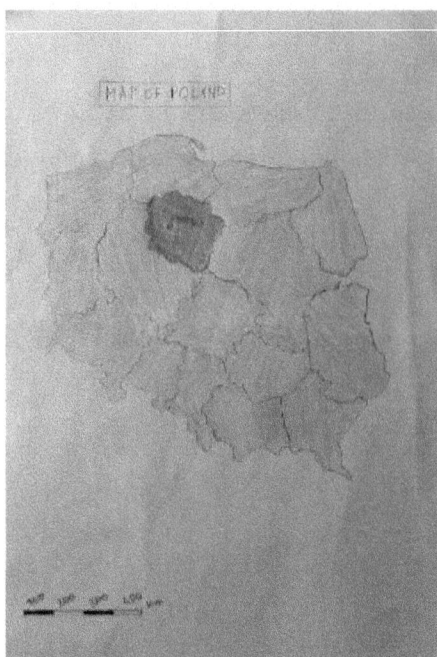

Figure 1, Map of Poland drawn by me

Legend of the map of Poland

Light blue areas	Regions in Poland except Kujawsko Pomorskie
Dark Blue areas	Kujawsko Pomorskie Region
Red Pointed area	Bydgoszcz

Before the research, I studied information about the UHI, which would help me to observe this phenomenon in a real life situation (the research I would do myself). As I started the research, I noticed a significant difference in the temperature in different areas of Bydgoszcz, which in other words shaped the phenomenon of the UHI in the city that I live in! This phenomenon as we can tell cannot be observed in any small town or city, it has to be a significantly big city that hosts thousands or millions of citizens and away from any forests and suburbs so the dissimilarity in the temperature is substantial.

My main aim is to find out if the city that I live in (Bydgoszcz) is affected by the UHI phenomenon. Furthermore, I am going to examine whether the variation in the temperatures are caused by the season, the time or the building density (FSI ratio[3])

STATEMENT AND JUSTIFICATION OF THE HYPOTHESIS

First hypothesis:

My first hypothesis is that the temperature in the center of the city will be higher than the temperature in the surrounding area, e.g. suburbs, parks, forest areas.

This would happen because the city center is far more populated, consequently there are more buildings supplied with home appliances, more cars and other resources that cover the human needs that emit heat

Second hypothesis:

My second hypothesis is that the temperature in the city is directly affected by the building density.

Its reasonable to say that the temperature in the city center would be higher, as more people live there, this means that there are more buildings for the citizens, more streetlamps are on during the night and, and more cars.

Third hypothesis:

My third hypothesis is that there would be a bigger difference in the temperature between first and last point of my measurement during winter.

This would be caused due to the building density that would help to maintain the heat between the buildings comparing to the suburbs or the forest areas that would be affected more by the low temperature. This bigger difference could be also caused by human activity in the city center.

---

[3] FSI is the ratio of building floor covered area to area available on the land.

121

In order to find out whether the city in live in, Bydgoszcz is effected by the urban heat island, I had to go out and investigate if there is a difference in the temperature between the center of the city and the suburbs. I came up with a plan of the measurement of the temperature.

Firstly, I had to measure the temperature every 250m. The first point that I measured the temperature was Szubińska Street. I was measuring the temperature every 250m. As I already mentioned the first measurement was at Szubińska Street and the last one was close to Stary Rynek (the old town of Bydgoszcz). The distance between my first point of measurement and my last point was approximately 2.7 km.

I decided to measure make my measurements 3 times per day (at 6:00 a.m., 12:00 p.m. and also at 7:00 p.m.) with the intention of better and more accurate results and examine whether there is a difference between morning, afternoon and evening since temperatures tend to fluctuate.

I measured the temperatures 2 times in different seasons, during winter and during spring in order to determine if the season affects the temperature change between the city center and the suburbs.

In order to have more precise and accurate measurements I used an electronic thermometer that was suggested from my teacher, in order to be easier for me to write down the different temperatures and avoid any random errors. I have listed my results down below according to the day, the season, the building density of the point, and the time.

For the purpose of this investigation, using the temperatures that I collected and the density of the buildings I used two different ways to find their correlation so I would be more certain about this relationship. I calculated these two coefficients using my electronic calculator. The two coefficients are given below each table.

As I had collected data for the temperature and the building density I used two methods to find the correlation. These correlations are called Pearson (rp) and Spearman (sp). Using both of these correlations helped me understand how the building density and the temperature in my city are connected, but also to ensure that it is understandable and well presented.

Spearman coefficient:

$$\rho = 1 - \frac{6\Sigma d_i^2}{n(n^2 - 1)}$$

Pearson coefficient:

$$r = \frac{n(\Sigma xy) - (\Sigma x)(\Sigma y)}{\sqrt{[n\Sigma x^2 - (\Sigma x)^2] \times [n\Sigma y^2 - (\Sigma y)^2]}}$$

Figure 2, shows the measurement points in Bydgoszcz, moving from the suburbs of the city towards the centre, moving from southwest to northeast.

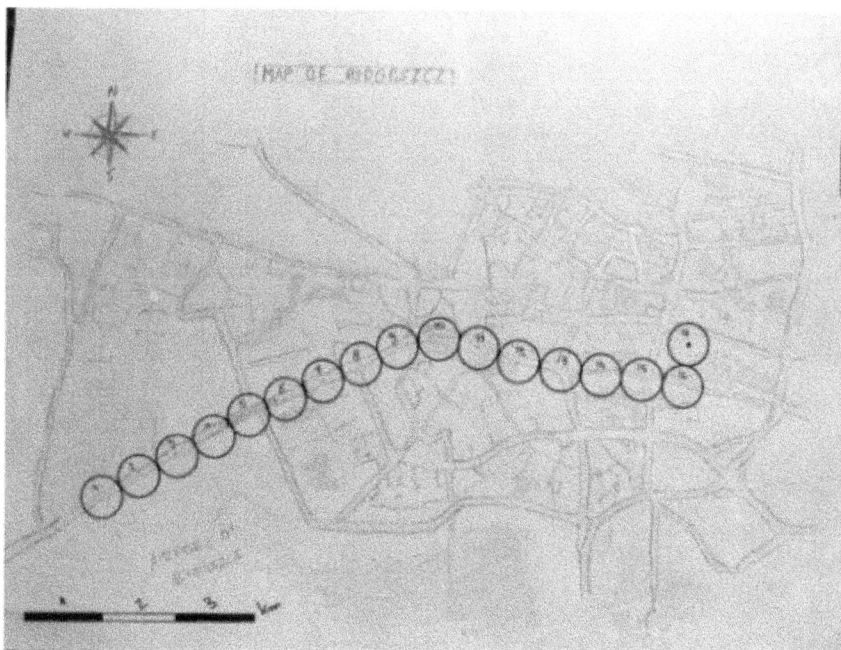

Figure 2, Map of Bydgoszcz presenting the stops at which I measured the temperature

123

Numbered Circles	Measuring points
Green areas	Parks or Courts
Yellow lines	Main Roads

This map demonstrates the route that we followed as well as the 17 points that we measured the temperature.

ANALYSIS

RAW DATA

Table 1, presentation of raw data, Pearson and Spearman coefficient.

stop number	Building density %	Temperatures on 27th of January			Temperatures on 29th of January			Temperatures on 19th of May			Temperatures on 21st of May		
		at 6.00	at 12.00	at 19.00	at 6.00	at 12.00	at 19.00	at 6.00	at 12.00	at 19.00	at 6.00	at 12.00	at 19.00
1	20	-7.8	-4.5	-4.2	-0.6	0.1	-7.5	13	16.8	15.4	15.4	24	18.7
2	30	-7.8	-4.4	-4.1	-0.6	0.1	-7.3	13	17.1	15.4	15.4	24.3	18.9
3	35	-7.8	-4.4	-4.1	-0.5	0.2	-7.1	13	17.1	15.6	15.6	24.5	19.1
4	40	-7.3	-4.2	-4	-0.5	0.2	-7	13.1	17.3	15.7	15.6	24.7	19.2
5	50	-7.2	-4.2	-3.8	-0.3	0.3	-6.5	13.1	17.5	15.9	15.6	24.7	19.3
6	60	-7	-4.1	-3.8	-0.2	0.3	-5.8	13.3	17.8	15.9	15.7	24.9	19.3
7	50	-7	-4	-3.7	-0.2	0.5	-5	13.5	17.9	15.9	15.7	25	19.4
8	50	-7	-3.8	-3.7	-0.1	0.5	-3.9	13.5	18.1	16.1	15.8	25.1	19.5
9	60	-6.8	-3.8	-3.5	-0.1	0.5	-3.8	13.7	18.2	16.3	15.9	25.3	19.6
10	40	-6.8	-3.7	-3.3	-0.1	0.6	-3.8	13.7	18.2	16.3	16	25.3	19.6
11	50	-6.8	-3.7	-3	-0.1	0.6	-3.3	13.9	18.3	16.5	16	25.3	19.6
12	65	-6.7	-3.5	-2.9	0	0.7	-3.2	13.9	18.4	16.5	16	25.4	19.7
13	70	-6	-3.3	-2.6	0	0.8	-3	14.2	18.4	16.7	16.1	25.4	19.7
14	80	-5.7	-2.4	-2.4	0	0.8	-2.2	14.2	18.5	16.7	16.1	25.5	19.8
15	90	-5.7	-2.3	-2.4	0.1	0.9	-2.2	14.2	18.7	16.9	16.2	25.6	19.8
16	100	-5.6	-2.2	-2	0.1	1	-2	14.5	18.7	17.1	16.3	25.6	19.8
17	100	-5.6	-2	-1.8	0.2	1.1	-2.4	14.5	18.7	17.1	16.3	25.6	19.8
pearson coefficient		0.947314	0.940375	0.923776	0.886085	0.914132	0.84111	0.899911	0.869481	0.921511	0.89076	0.854559	0.848108
spearman coefficient		0.927103	0.900324	0.906516	0.921928	0.896276	0.907875	0.905978	0.915639	0.91491	0.888612	0.914856	0.910794

GRAPHED DATA

Temperature 27.01.21

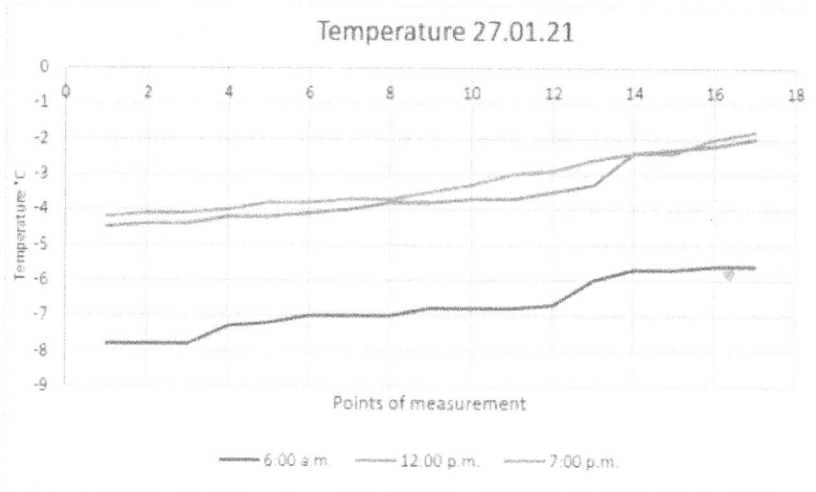

Figure 3, Graph representing the temperature for each stop on 27ᵗʰ of January

Temperature on the 29.01.2021

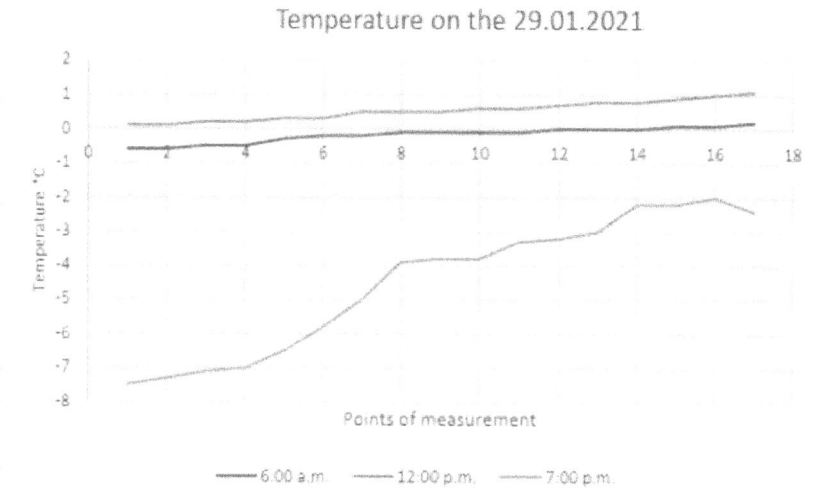

Figure 4, Graph representing the temperature for each stop on 29ᵗʰ of January

Temperature on the 19.05.21

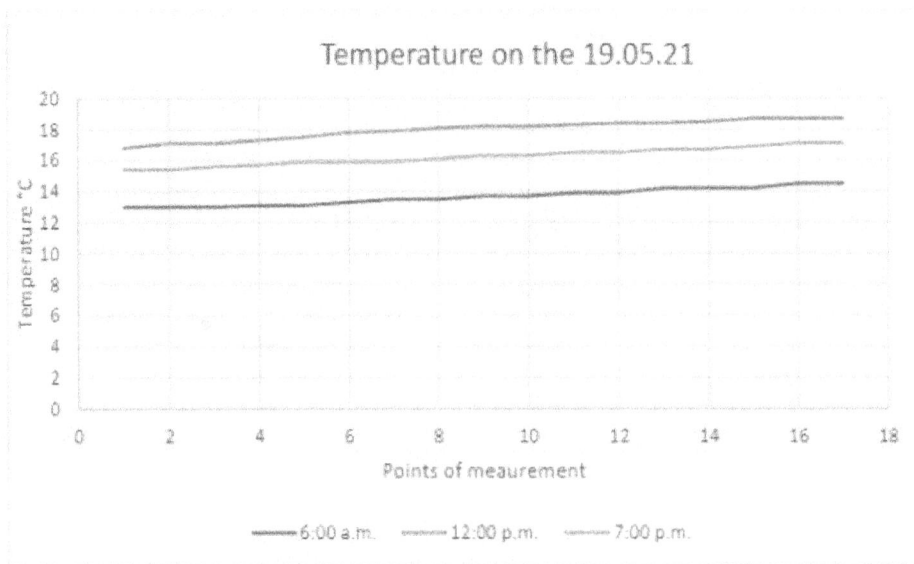

Figure 5, Graph representing the temperature for each stop on 19th of May

Temperature on the 21.05.21

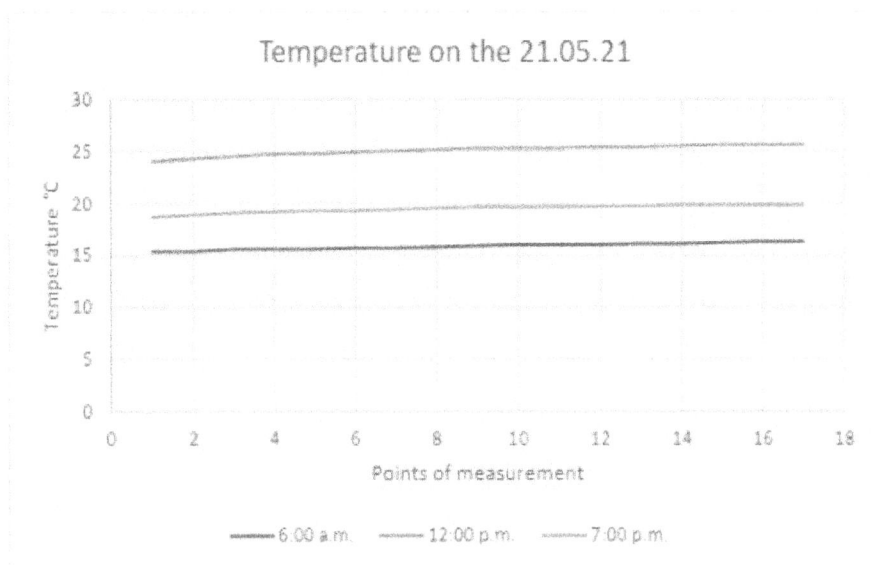

Figure 6, Graph representing the temperature for each stop on 21st of May

BUILDING DENSITY

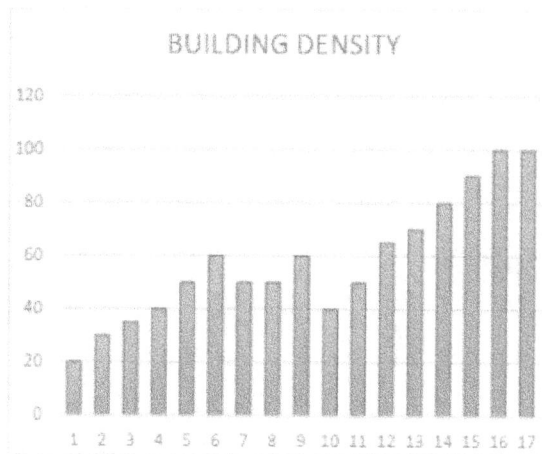

Figure 7, graph representing the building density for each stop

The graph above displays how the building density is increasing as we moving from the suburb and the forest areas away from the city, towards the center of the city. In general, it can be stated that there is a notable increase in the building density and surely, this is one of the factors affecting the temperature difference in my measurements.

Pearson coefficient values

Figure 8, graph presenting the Pearson coefficient values from Table 3

Spearman coefficient values

Figure 9, ,graph presenting the Spearman coefficient values from Table3

DISCUSSION

Observing **Figure 3**, we can say that there is a significant change in temperature comparing the first point to the last point, in other words moving towards the center of the city. Noteworthy is also the average change in temperature from the first to the last point, which is 2.36°C

Looking carefully into **Figure 4** we can see a considerable increase in the temperature comparing to the measurement on the 29th of January. Especially for the measurement at 6:00 am, as the difference in temperature is 5.1°C. Once more there is a remarkable change as we move to the city center (1st to 17th point). The average change of temperature is 2.33°C

Figure 5 represents how does the temperature change as we move closer to the city center during spring, and more specifically on 19.05.2021. It is apparent that the temperature increased comparing to the winter, and the difference of temperature of the suburbs of the city and the center of the city is not as big.

In **Figure 6**, it is obvious that the temperature is not fluctuating as it has an average temperature change of only 1.2 °C.

From the graph of the Pearson coefficient (**Figure 8**), and from Table 3, we could say that the values are distributed irregularly, as they vary from 0.84111 to 0.947314. Although,

128

the Spearman coefficient (**Figure 9**) shows us that the temperature increases for each stop, with the lowest value of 0.888612 and the highest of 0.927103.

CONCLUSION

Concisely, by carefully observing every measurement and determining the average difference of temperature in each day, it is proven that the Urban Heat Island effect occurs in Bydgoszcz, as the temperatures outside of the city center (suburbs) are significantly lower than the temperatures that occur in the center of the city. This also proves that my **first hypothesis**, that the temperature in the city center would be higher due to traffic, and home appliances, is correct.

I believe that the temperature of an area is connected with the building density , as we can observe throughout the whole investigation getting closer to the center of Bydgoszcz leads to much higher temperatures due to the ability of the buildings to absorb and maintain heat, therefore that proves that the building density truly affects the temperature of the city. This relation between the increase of the building density and the temperature proves that my **second hypothesis** is correct. Although this can be wrong sometimes, as there are many factors that could alter the pattern that I observed (high building density-higher temperatures). Factors such as rain, snow could result in lower temperatures in the center of the city than the suburbs.

Additionally, by comparing the average change of temperature for both winter measurements (27th and 29th of January): 2.33°C and spring (19th and 21st) of May 1.45°C) we can say that there is a bigger change in temperature for winter as we move closer to the city center and the building density increases. Which confirms my **third hypothesis**.

Overall, it was a fun experience for me as it helped me widen my knowledge on the phenomenon of urban heat island. Even though I faced some challenges, I was able to cope with them and overcome them. It also helped me gain experience on how to conduct a research, how to investigate and present my data, which for sure will help me with my future studies.

EVALUATION

I am satisfied and pleased with my investigation and I believe that my fieldwork is dependable. Surely, this investigation can be improved and get more precise and accurate results. This could happen by Measuring the measurement with a more precise and professional thermometer so as to avoid any uncertainties and random errors. Moreover, another way of getting closer to the ideal results would be to add even more data, at least 3 more days for each season (winter and spring). This would give us a wider range of temperatures and would result in a more valid investigation. My investigation required to measure the temperature several times in a day, so that lead to spending time on moving from one point to another. Spending time to move from point to point means that it was not possible to measure all 17 points at the same time, which would be the ideal way for my investigation.

On the other hand, I would like to emphasize the cons of my research and what I think was well terminated.

As I mentioned before, the use of an electronic thermometer for sure allowed me to get precise measurements, without making any reading errors.

Measuring the temperature 3 times in a day for sure clarified the difference in temperature between morning, afternoon, and evening. Likewise, measuring the temperature in two different seasons for sure proved that the season is also one of the factors that are responsible for the Urban Heat Island phenomenon.

Lastly, measuring the temperature with an electronic thermometer but also measuring the temperature in the same exact spots each day definitely enhanced the reliability of my results.

BIBLIOGRAPHY + APPENDICES OMITTED

5. EXAMPLE FIVE (25/25)

Name To what extent does the city centre of Katowice,

Poland, fit the central business district (CBD) model?

Author: Joanna Piotrowska

Session: May 2023

Level: SL

1. Introduction

1.1. Geographical context

Fieldwork took place in Katowice city, where our school is located, hence everyone in the class is more or less familiar with the area, which is easily accessible, and so everyone could participate in data collection. Within the group we agreed that it would be interesting to see the city we go through almost every day from a different, more scientific perspective.

Katowice attained city status in 1865 and were developing rapidly due to industrial growth. Now it is capital city of Upper Silesian Voivodeship in south Poland[1]. It has an area of 165 km^2 (12th greatest city in Poland)[2] and is inhabited by 287,000 people (11th greatest population in Poland)[3]. Transport system is well-developed, meaning public transport within the region, as well as intercity and international buses, railways and airport. The city provides good educational and cultural opportunities and is a retail and commercial centre[4]. Subregion of Katowice has a large share in Polish GDP, in 2019 it amounted to 2.6%[5].

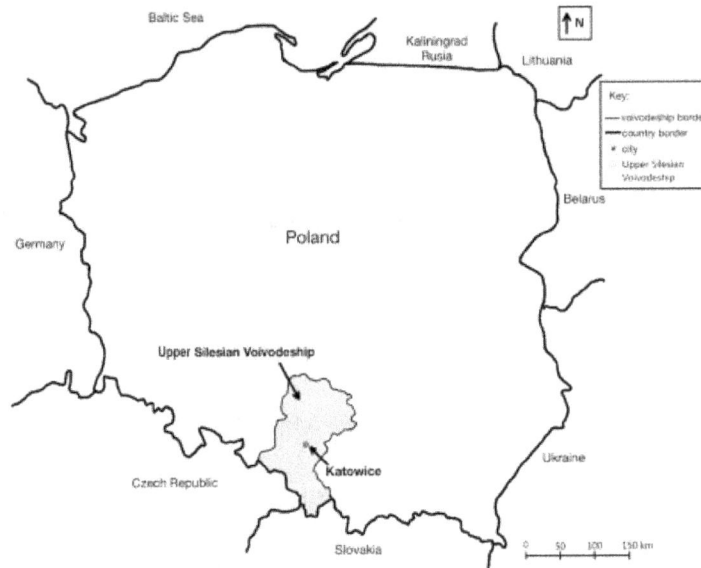

Figure 1 Map of Poland - Silesian Upper Voivodeship and Katowice city showed[6].

---

[1] *Welcome to Katowice*, https://welcome.katowice.eu/ [access: 08.06.2022].

[2] *Miasta o największej powierzchni w Polsce*,
https://www.polskawliczbach.pl/miasta_o_najwiekszej_powierzchni_w_polsce [access: 09.06.2022].

[3] *Największe miasta w Polsce pod względem liczby ludności – ranking TOP 15*, https://www.national-geographic.pl/artykul/najwieksze-miasta-w-polsce-gdzie-mieszka-najwiecej-ludzi [access: 09.06.2022].

[4] *Welcome to Katowice*, https://welcome.katowice.eu/ [access: 08.06.2022].

[5] A. Hetmańska, *Produkt krajowy brutto w regionie śląskim w 2019 r.*, https://katowice.stat.gov.pl/opracowania-biezace/opracowania-sygnalne/rachunki-regionalne/produkt-krajowy-brutto-w-regionie-slaskim-w-2019-r-,1,17.html [access: 08.06.2022].

[6] Candidates own drawing, based on: *Mapa Polski*, https://mundosklep.pl/pl/strona-glowna/98169-mapa-polski-65-x-50-cm-wojewodztwa-i-ich-stolice-mapa-do-kolorowania.html [access: 08.06.2022].

1.2. Fieldwork question

To what extent does the city centre of Katowice, Poland, fit the central business district (CBD) model, based on the five characteristic features of CBD?

1.3. Hypothesis

Visible features of big, important culturally and economically city suggest that Katowice city centre will fit the CBD model and hence it will possess five characteristic features of CBD:

– the ratio of shops to other properties on the ground floor will be greater than $\frac{1}{3}$,

– the ratio of offices to other properties on the ground floor will be greater than $\frac{1}{10}$,

– buildings will be high and their height will decrease with increasing distance from the centre,

– land value will be greatest in the core of CBD,

– shops and offices will be clustered[7],

 to verify this hypothesis more detailed investigation of specific CBD characteristics was performed.

1.4. Methodology

Fieldwork data were collected by six members of geography class in our school. Four people collected data concerning type of land use and buildings' heights along the fragments of streets showed on the map below (Figure 2), one person collected data concerning relation of land value to distance from the city centre and one person collected data concerning clustering of shops and offices.

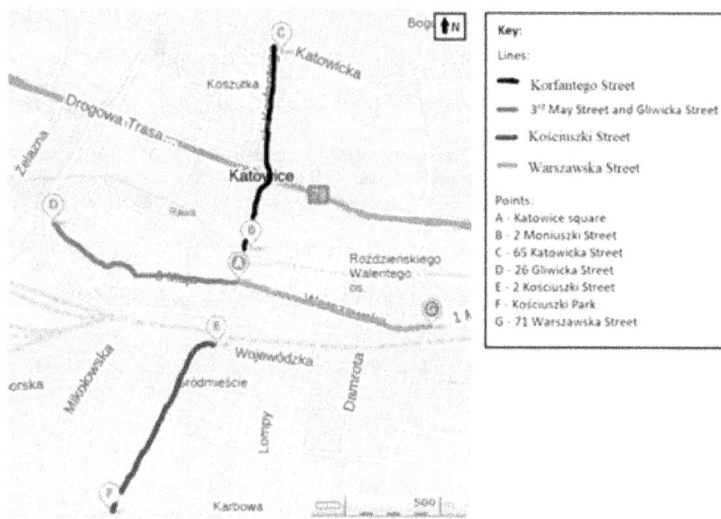

Figure 2 Korfantego, 3rd May, Gliwicka, Kościuszki, Warszawska streets' segments under investigation[8].

---

[7] D. Waugh, *Geography. An Integrated Approach*, 4th edition, Cheltenham: Nelson Thornes 2009, p. 430.

[8] Based on: *Katowice na mapie Targeo*, https://mapa.targeo.pl/katowice/miasta [access: 06.06.2022].

1. Segments of streets for investigation (Kościuszki, Warszawska, Korfantego, 3rd May and Gliwicka) were chosen by spatial, systematic sampling, as each of them is directed to different world's side from Katowice city centre. Each segment was approximately of the same length: Kościuszki – 1.4 km, Warszawska – 1.3 km, Korfantego – 1.2 km, 3rd May and Gliwicka – 1.1 km.

Table 1 Types of land uses and signs used to denote them.

Sign	Description
A	retailing
A_1	low order shops (e. g. food shops, bakeries)
A_2	middle order shops (e.g. chemists, clothes shops)
A_3	high order shops (e. g. wedding accessories shops, car shops)
B	restaurants, cafes, entertainment
C	offices, financial institutions (e. g. lawyers, banks)
D	other services (e. g. doctors, cobblers)
E	residential buildings
F	public buildings (e. g. schools, hospitals)
G	open space (e. g. parks, lawns)

2. Type of land use (Table 1) data were collected from observation along the streets mentioned above. Only land uses located on ground floor were taken into consideration, if more than one type of land use was present on the ground floor of the building, all of them were counted.

3. Buildings' heights data were collected from observation along the same streets. Height was measured in floors with ground floor and attic counted as 1 floor.

4. To determine relation between land value and distance from the city centre, on each of the abovementioned streets 3 apartments for sale were found through online advertisements in services *Nieruchomości-online.pl*, *SonarHome.pl* and *Gratka.pl*. They were chosen by convenience sampling as only few offers were available. All the apartments were two-room, their area was 49 ± 11 m^2 and distance from the city centre varied from 350 m to 3.9 km. Information about the state of the apartment and price per m^2 was collected.

5. To determine if offices and shops are clustered nearest neighbour index (NNI) was calculated from the formula

$$\text{NNI} = 2\bar{D}\sqrt{\frac{N}{A}},$$

Figure 3 Formula used for NNI calculation[9].

where:

$\bar{D}$ – average distance between each point and another point closest to it,

N – number of points under investigation,

A – size of the area under investigation.

[9] G. Naggle, B. Cooke, *Skills for IB Geography*, Oxford: University Press 2017, www.oxfordsecondary. co.uk/9780198396031 [access: 02.01.2023].

NNI allows to distinguish three patterns: clustered, regular and random. Its results are within [0, 2.15] interval, where 0 means clustered, 1.0 – random and 2.15 – regular[10].

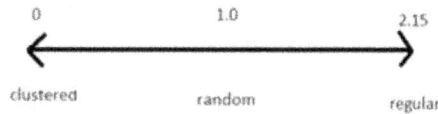

0   1.0   2.15

clustered   random   regular

Figure 4 Tendencies toward clustered and regular distribution according to NNI values[11].

Data needed for calculating NNI (distances between each office and shop and another office or shop closest to it, number of offices and shops under investigation) were collected from the area within 1 km radius from city centre, using online map Targeo.

1.5. Link to the syllabus

Fieldwork is strictly connected to Option G – *Urban environments*, especially topic – *The variety of urban environments*, in which we discussed model of CBD and its features, got to know advantages and problems of big cities and explored some case studies. Knowledge gained in this unit allows to compare features of Katowice city centre with CBD model and hence answer the research question for this fieldwork.

2. Data analysis

CBD is the main part of the city, characterized by high offices, retailing and public transport concentration. Buildings there grow upwards due to high land values and vertical zoning often occurs with shops on ground floors and offices on upper levels. Similar types of land uses are often clustered in the same area[12].

2.1. Land use

As mentioned above, land use in CBD is characterized by clustering of similar types of land use in the same area. From visualised data of land use on four chosen streets (Figure 5) examples of this clustering can be found, such as retailing concentration on 3rd May Street and north part of Kościuszki Street or residential buildings concentration in south part of Kościuszki Street.

---

[10] *Nearest Neighbor,* https://sites.google.com/site/geographyfais/fieldwork/6-data-analysis/statistical-tools/clustering-dispersal/nearest-neighbor [access: 06.06.2022].

[11] Candidates own diagram based on: *Nearest Neighbor,* https://sites.google.com/site/geographyfais/fieldwork/6-data-analysis/statistical-tools/clustering-dispersal/nearest-neighbor [access: 06.06.2022].

[12] G. Nagle, B. Cooke, *Geography. Course Companion,* 2nd edition, Oxford: University Press 2017, p. 340.

Figure 5 Map of investigated streets showing types of land uses[13].

As stated before CBD is characterized by high retailing concentration. The ratio of shops to other properties on ground floor in the CBD should be higher than $\frac{1}{3}$. Number of shops and other properties on the investigated streets are shown in the Table 2 together with the percentage that shops constitute of all the buildings.

[13] Candidate's own drawing based on: *Katowice na mapie Targeo*, https://mapa.targeo.pl/katowice/miasta [access: 12.06.2022].

Table 2 Number and percentage of shops, number of other properties.

Street name	Korfantego	Gliwicka & 3rd May	Kościuszki	Warszawska	Overall
Number of shops	23	51	37	16	127
Number of other properties	41	76	63	63	243
Percentage of shops [%]	36	40	37	20	35

On the graph below (Figure 6) percentages of shops and other properties on each of the investigated streets and overall are shown with baseline at 33.33%.

Figure 6 Percentage of shops and other properties on ground floor.

Except for Warszawska street, ratio of shops to other properties on ground floor of other investigated streets exceeds the ratio of $\frac{1}{3}$. The inconsistency of ratio on Warszawska street with the assumption might be caused by high number of old tenement houses[14], which are mostly used as offices and residential buildings, which can be concluded from the map showing types of land use (Figure 5). Overall the ratio of shops to other properties on investigated streets exceeds $\frac{1}{3}$, which is consistent with the hypothesis.

Although clustering of offices is another feature of CBD, the minimal ratio of offices to other properties on ground floor is smaller than in case of shops, it should be higher than $\frac{1}{10}$. This is due to vertical zoning, which indicates that offices will be placed on higher floors. Number of offices and other

---

[14] *Wartości dziedzictwa kulturowego,*
http://web.archive.org/web/20160304134848/http://bip.um.katowice.pl/dokumenty/2010/1/5/1262697057.pdf [access: 06.01.2023].

properties on the investigated streets are shown in the Table 3 together with the percentage that offices constitute of all the buildings.

Table 3 Number and percentage of offices, number of other properties.

Street name	Korfantego	Gliwicka & 3rd May	Kościuszki	Warszawska	Overall
Number of offices	11	18	5	10	44
Number of other properties	53	109	95	69	326
Percentage of offices [%]	17	14	5	13	12

On the graph below (Figure 7) percentages of offices and other properties on each of the investigated streets and overall are shown with baseline at 10%.

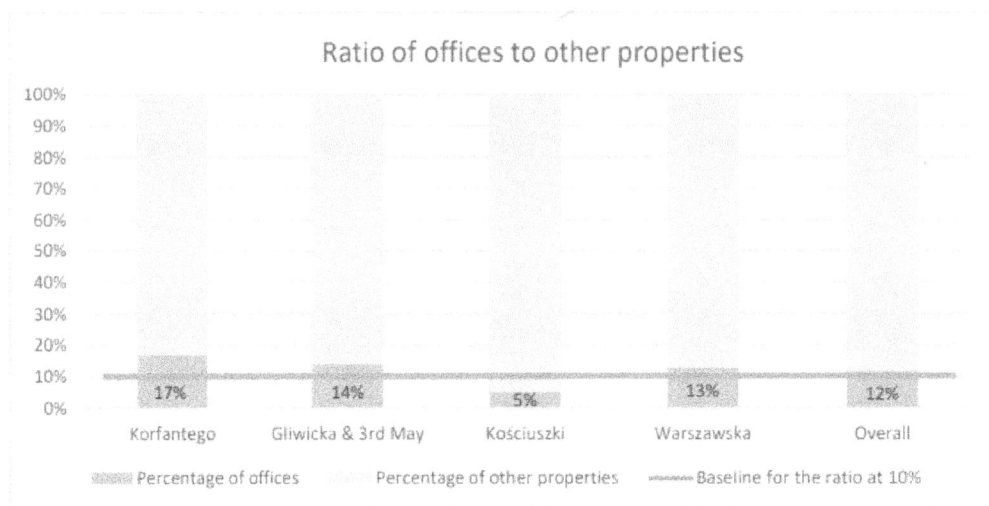

Figure 7 Percentages of offices and other properties on ground floor.

Except for Kościuszki Street, the ratio of offices to other properties on ground floor of other investigated streets exceeds $\frac{1}{10}$. Lower ratio on Kościuszki Street might be caused by clustering of shops in the north part of the street and clustering of residential buildings in the south part of the street (Figure 5), which doesn't leave space for high concentration of offices there. Overall the ratio of offices to other properties on investigated streets exceeds $\frac{1}{10}$, which is consistent with the hypothesis.

2.2. Height of the buildings

In CBD buildings grow vertically due to high land prices. Buildings' height will be generally big and according to CBD model it should decrease with distance from the centre[15].

---

[15] D. Waugh, *Geography. An Integrated Approach*, 4th edition, Cheltenham: Nelson Thornes 2009, p. 430.

Table 4 Mean heights of buildings on investigated fragmets of streets.

Street name	Korfantego	Gliwicka & 3rd May	Kościuszki	Warszawska
Mean height [floors]	5.81	4.28	3.98	3.63

Buildings in Poland are classified into 4 categories: low (up to 4 floors), medium-high (5-9 floors), high (10-18 floors) and skyscrapers (over 55 m high)[16]. Collected data show that buildings in city centre of Katowice generally can be considered as low or medium-high, but there are some buildings which can be classified as high (Figure 8). However, most of the buildings commissioned between 2005 and 2019 have only 2 floors, hence mean heights on investigated streets are greater in comparison to Polish average[17].

On the graphs below (Figures 8, 9, 10, 11) height of buildings in order of increasing distance from centre on each of the investigated streets is shown.

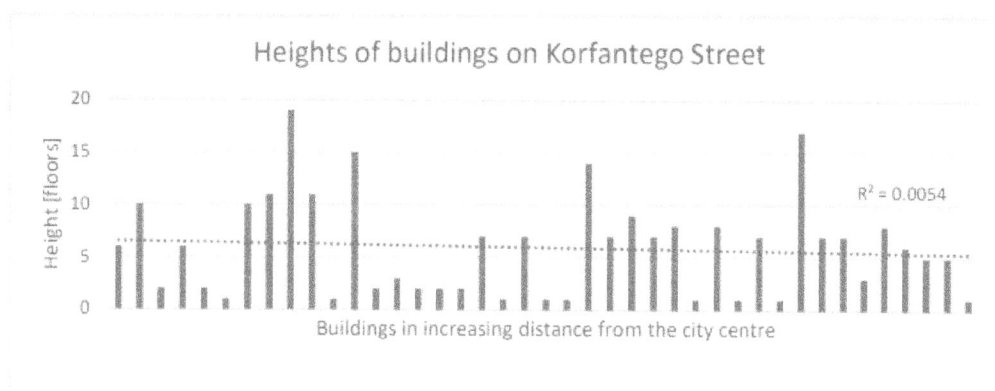

Figure 8 Buildings' heights on Korfantego Street in increasing distance from centre.

---

[16] M. Pol, *Rozporządzenie ministra infrastruktury z dnia 12 kwietnia 2002 r. w sprawie warunków technicznych, jakim powinny odpowiadać budynki i ich usytuowanie,*
https://isap.sejm.gov.pl/isap.nsf/download.xsp/WDU20020750690/O/D20020690.pdf [access: 12.01.2023].
[17] A. Dobkowska, *Ile kondygnacji mają budynki mieszkalne budowane w Polsce?*, https://www.locja.pl/raport-rynkowy/ile-kondygnacji-maja-budynki-mieszkalne-budowane-w-polsce,192 [access: 12.01.2023].

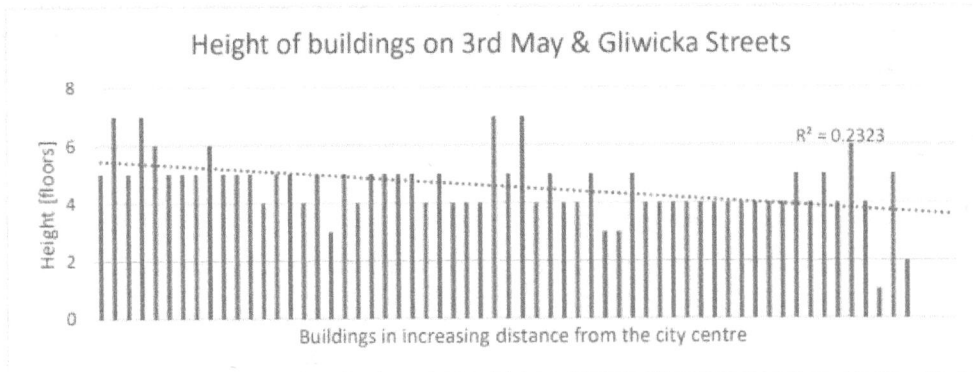

Figure 9 Buildings' heights on 3rd May and Gliwicka Streets in increasing distance from centre.

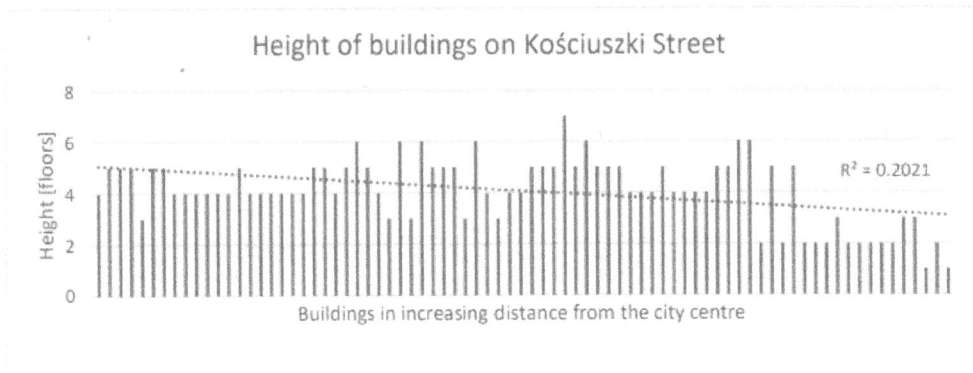

Figure 10 Buildings' heights on Kościuszki Street in increasing distance from centre.

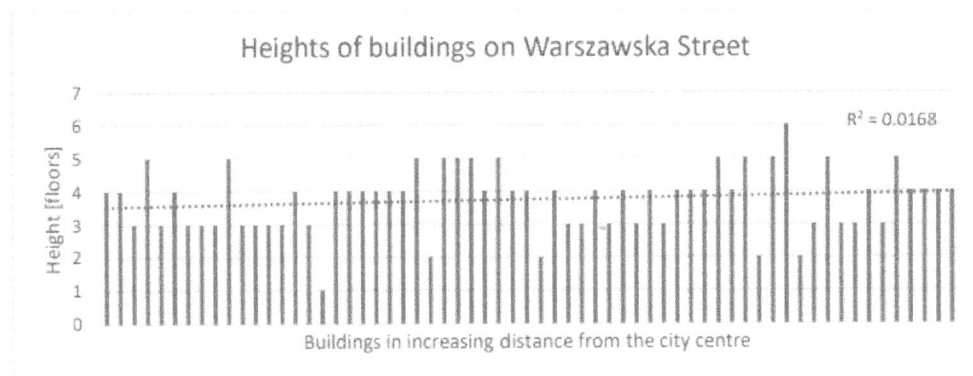

Figure 11 Buildings' heights on Warszawska Street in increasing distance from centre.

In case of Korfantego and Warszawska streets, R^2 values of 0.0054 and 0.0168 respectively indicate that regression model applied don't fit the data. Hence, although linear trendline shows slightly negative correlation in case of Korfantego street and slightly positive correlation in case of Warszawska street, they cannot be considered valid. In case of 3rd May with Gliwicka streets and Kościuszki street the R^2 values of 0.2323 and 0.2021 respectively indicate that regression model applied weakly fit the data. Hence, the negative trend supports the hypothesis, that building height will decrease with distance from the centre, but the result cannot be considered certain.

The lack of or weak correlation between height of buildings and distance from the city centre and hence inconsistency with the CBD model could be explained by non-uniform architectonic style[18] and many historical buildings in the centre of Katowice[19] which doesn't allow for construction of new tall buildings. High offices and residential buildings are built in free spaces or further away from the centre.[20]

2.3. Land values

Table 5 Land values and factors affecting prices.

Street	No	Distance from city centre [m]	Meterage [m²]	Number of rooms	Status [ready to live in/for renovation]	Price [zł/m²]	Source
Korfantego	1	450	53	2	for renovation	6 584.91	21
	2	600	48	2	ready to live in	6 458.33	22
	3	600	49	2	ready to live in	7 326.53	23
Gliwicka	1	1800	56	2	for renovation	5 892.86	24
	2	1900	45	2	ready to live in	7 977.78	25
	3	3900	39	2	ready to live in	7 970.08	26
Kościuszki	1	350	58	2	for renovation	7 241.38	27
	2	950	48	2	for renovation	7 950.00	28
	3	1200	38	2	ready to live in	7 881.58	29
Warszawska	1	450	60	2	no data	5 055.00	30
	2	500	59	2	ready to live in	6 777.97	31
	3	550	56	2	ready to live in	7 113.00	32

[18] *Katowickie Perełki*, https://slaskiekamienice.pl/2021/06/11/katowickie-perelki/?fbclid=IwAR1rQf8Yf6VhkfGEOwJ-bUDes3aduDPTvgYYoZ4YoUzVrDma51RVKyciLVY [access: 12.01.2023].

[19] *Wartości dziedzictwa kulturowego*,
http://web.archive.org/web/20160304134848/http://bip.um.katowice.pl/dokumenty/2010/1/5/1262697057.pdf [access: 06.01.2023].

[20] *Inwestycje Katowice*, https://investmap.pl/miasto/katowice [access: 12.01.2023].

[21] https://katowice.nieruchomosci-online.pl/mieszkanie-w-bloku-mieszkalnym,do-remontu/23153309.html [access: 09.05.2022].

[22] https://katowice.nieruchomosci-online.pl/mieszkanie-w-bloku-mieszkalnym,do-odswiezenia/23197955.html [access: 09.05.2022].

[23] https://katowice.nieruchomosci-online.pl/mieszkanie,z-oddzielna-kuchnia/22680815.html [access: 09.05.2022].

[24] https://katowice.nieruchomosci-online.pl/mieszkanie,m3,z-oddzielna-kuchnia/23318906.html [access: 09.05.2022].

[25] https://katowice.nieruchomosci-online.pl/mieszkanie,z-kuchnia-z-oknem/23162592.html [access: 09.05.2022].

[26] https://katowice.nieruchomosci-online.pl/mieszkanie,z-kuchnia-z-oknem/23222157.html [accessed: 09.05.2022].

[27] https://katowice.nieruchomosci-online.pl/mieszkanie,m2,z-kuchnia-z-oknem/23319304.html [accessed: 09.05.2022].

[28] https://katowice.nieruchomosci-online.pl/mieszkanie,m2,z-oddzielna-kuchnia/23087322.html [accessed: 09.05.2022].

[29] https://katowice.nieruchomosci-online.pl/mieszkanie,m2,z-aneksem-kuchennym/22472086.html [accessed: 09.05.2022].

[30] https://sonarhome.pl/ceny-mieszkan/katowice/srodmiescie/warszawska/25/ma85gy7u [accessed: 09.05.2022].

[31] https://gratka.pl/nieruchomosci/mieszkanie-katowice-srodmiescie-ul-warszawska/ob/22817897 [accessed: 09.05.2022].

[32] https://katowice.nieruchomosci-online.pl/szukaj.html?3,mieszkanie,sprzedaz,,Katowice:19572,,Warszawska:193299,,,-70 [accessed: 09.05.2022].

Table 6 Mean prices in Katowice city centre, Katowice city and Poland.

Mean price [zł/m²] in Katowice city centre district	7019
Mean price [zł/m²] in Katowice	6314[33]
Mean price [zł/m²] in Poland in I quarter of 2022	5252
Mean price [zł/m²] in Poland in II quarter of 2022	5020
Mean price [zł/m²] in Poland in III quarter of 2022	5295[34]

Mean price of m² in Katowice city centre district calculated from the data collected (Table 6) is 7019 zł, which is higher than mean in whole Katowice city (6314 zł) as well as mean price in Poland (5252, 5020, 5295 zł respectively in I, II and III quarter of 2022).

As status of apartment is important factor influencing its price, prices of apartments with status ready to live in and for renovation are plotted separately against distance from the city centre (Figures 12 and 13).

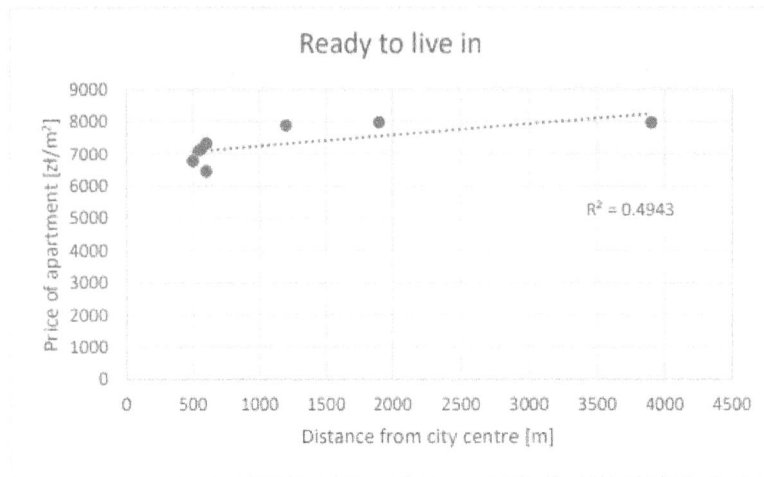

Figure 12 Relation between ready to live in apartments' prices and distance from centre.

In the graph above positive correlation between price per m² of apartments with status ready to live in and distance from the city centre can be observed. R^2 value indicates that regression model moderately fits the data. The positive correlation might be due to old tenements in the centre and newly built residential buildings further away[35]. It corresponds also to the suburbanisation process occuring in Katowice[36].

---

[33] Ceny mieszkań: Katowice, https://sonarhome.pl/ceny-mieszkan/katowice?fbclid=IwAR0APQrC3Si5VHJBFxPUZeyfTpE9dL5tJip8Txbpbp5DaFygA3ax_O3-io4 [access: 07.06.2022].
[34] Cena 1 m2 powierzchni użytkowej budynku mieszkalnego oddanego do użytkowania, https://stat.gov.pl/obszary-tematyczne/przemysl-budownictwo-srodki-trwale/budownictwo/cena-1-m2-powierzchni-uzytkowej-budynku-mieszkalnego-oddanego-do-uzytkowania,8,1.html [access: 13.01.2023].
[35] Inwestycje Katowice, https://investmap.pl/miasto/katowice [access: 12.01.2023].
[36] T. Spórna, The suburbanisation process in a depopulation context in the Katowice conurbation, Poland, 'Environmental & Socio-economic Studies' 2018, vol. 6, ISS. 1, pp. 57-72, doi: 10.2478/environ-2018-0007 [access:14.01.2023].

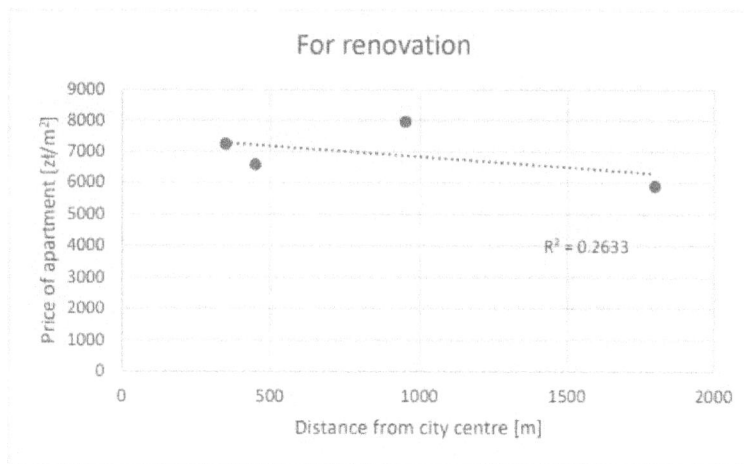

Figure 13 Relation between for renovation apartments' prices and distance from centre.

In the graph above negative correlation between price per m² of apartments with status for renovation and distance from city centre can be observed. However, R^2 value indicates that regression model weakly fits the data. Another factor decreasing reliability of the result is low number of apartments taken into consideration. Nevertheless price of apartments with status for renovation might be better indicator of land value than price of apartments with status ready to live in as no additional advances influence the price. In this case the negative correlation fits the hypothesis that land value will be the highest in the very centre.

2.4. Clustering

In order to determine if offices and shops are clustered in Katowice city centre, NNI was calculated from data collected: number of shops and offices, mean of distances between each shop and office to another shop or office closest to it within area of a circle with 1 km radius from city centre.

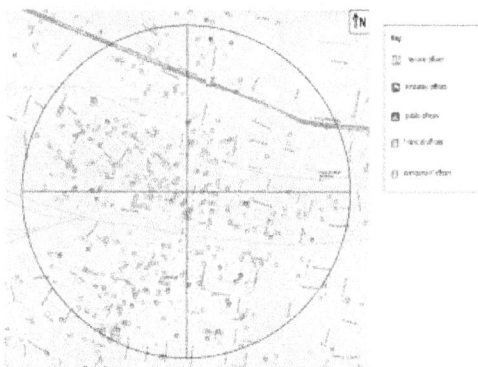

Figure 14 Area for NNI of offices calculations with all included points[37].

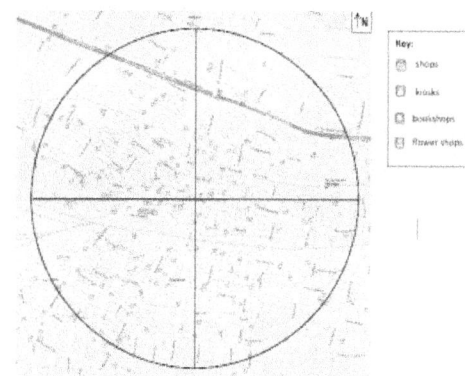

Figure 15 Area for NNI of shops calculations with all included points[38].

[37] Based on: *Katowice na mapie Targeo*, https://mapa.targeo.pl/katowice/miasta [access: 06.06.2022].
[38] Based on: *Katowice na mapie Targeo*, https://mapa.targeo.pl/katowice/miasta [access: 06.06.2022].

143

Table 7 Data needed for NNI calculations.

Number of shops in investigated area	149
Mean of distances between each shop and another shop closest to it [m]	65.6
Number of offices in the investigated area	301
Mean of distances between each office and another office closest to it [m]	38.8
Radius of a circle designating the area under investigation measured from the city centre [m]	1000

NNI for shops and offices was calculated using the formula mentioned in methodology (Figure 3).

$$NNI_{offices} = 2 \times 38.8 \sqrt{\frac{301}{\pi \times 1000^2}} = 0.76$$

$$NNI_{shop} = 2 \times 65.6 \sqrt{\frac{149}{\pi \times 1000^2}} = 0.90$$

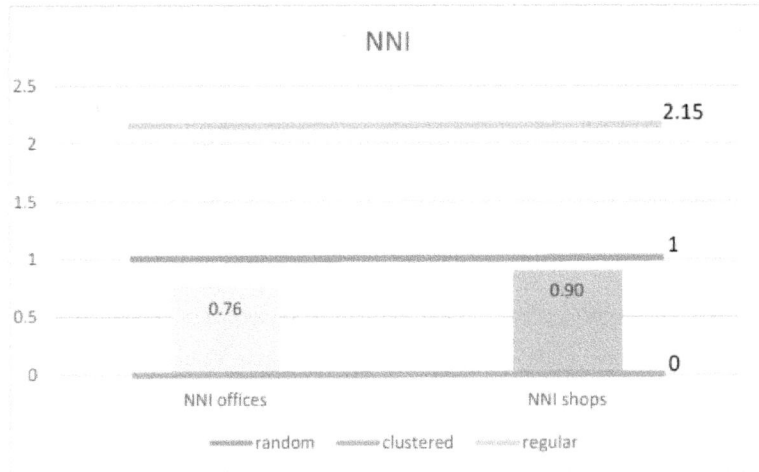

Figure 16 NNI of shops and offices, baselines at 0, 1, 2.15.

On the graph above calculated values of NNI for offices and shops are shown. Value of NNI for offices suggests that they are more clustered than shops, however both values are closer to 1 than 0, which implies random distribution.

In the map showing distribution of offices (Figure 14), much more clustered distribution can be noticed on the west side. This observation is justified by the characteristic feature of CBD, concerning functional segregation of land.

In the map showing distribution of shops (Figure 15), it can be observed that buildings are clustered near the centre, but dispersed in the peripheries. If the smaller area from the centre was chosen for investigation, the value of NNI would most probably indicate clustered distribution, which would support the hypothesis that shops will be clustered in the CBD.

3. Evaluation

3.1. Weaknesses and improvements

Table 8 Weaknesses, limitations, their influence on the results and possible improvements.

Weaknesses/limitations	Influence on the results	Improvements
many people were involved in data collection	some differences in method could exist	only one person chosen for one type of data collection
only 4 streets were chosen for investigation of land uses and buildings' heights	less reliable results due to small number of trials	investigate given features on bigger number of streets
meterage of apartments chosen for investigation of land values was different	the prices of apartments may differ due to differences in meterage	choose streets were apartments with same meterage can be found or choose any apartment with proper meterage at given distance from the city centre
small number of apartments chosen for investigation of land value	decreased reliability of obtained trend in relation of apartment's price to distance from the centre	find greater number of apartments for investigation by changing sampling method
NNI was calculated from data from online map	new buildings could have been built so calculated NNI would differ from real one	collect data on your own during field work
too big distance chosen for NNI calculation	NNI of shops and offices suggests more random than clustered distribution	choose smaller area for NNI calculation

3.2. Strengths

The investigation has some weaknesses and limitations, together with possible improvements, which should be considered during next field work. However, it has also several strong sides, including:

- relatively easy method, not requiring specific equipment;

- types of land use were determined in the field work, so data are up-to-date;

- investigated area was symmetrically distributed from the city centre.

3.3 Possible extensions

Different features of CBD model could be investigated:

- flow of pedestrians and vehicles could be measured at given points and time in different distances from the centre,

- concentration of public transport could be compared at different distances from the centre[39].

---

[39] D. Waugh, *Geography. An Integrated Approach*, 4th edition, Cheltenham: Nelson Thornes 2009, p. 430.

4. Conclusion

Land use in the area of investigation generally is consistent with CBD characteristics. Ratio of shops to other properties on most of the streets and overall exceeded $\frac{1}{3}$ and ratio of offices to other properties exceeded $\frac{1}{10}$. On the maps it could be seen that shops are clustered in the centre and that offices clustering is an example of functional segregation, however it was not confirmed by NNI calculations, which implied random distribution. Buildings were relatively high, but there was no significant correlation between building heights and distance from the city centre. Land value in the centre is high in relation to total Katowice city and Poland area. Prices of apartments with status for renovation decreased with increasing distance from the centre, however relatively small R^2 value, small number of apartments taken under investigation and opposite trend in case of apartments with status ready to live in reduced reliability of this result. Some incompatibilities of the results with the CBD model were caused by errors in method, for example too big area chosen for NNI calculations, which could be easily improved. Overall, after consideration of the results, errors and possible improvements for the investigation, despite some inconsistencies with the CBD model, the hypothesis that Katowice city centre fits the CBD model is supported by the investigation.

BIBLIOGRAPHY + APPENDICES OMITTED

6. EXAMPLE SIX (22/25)

Title: To what extent does river Hololo follow the Bradshaw model?

Author: B. Mehta

Session: May 2022

Level: SL

Examiner's summary

Criterion A [2/3]:

The student has successfully formulated a narrowly focused geographical fieldwork question and justified their choice of location. They collected primary and secondary data and analyzed it using the Bradshaw model. The presented geographical theory is well explained and includes all necessary background information. Additionally, the student has identified and explained the link between their study and the areas of the geography syllabus. However, the locational maps provided are unclear and need improvement.

Criterion B [2/3]:

The student demonstrated a strong understanding of the methods selected for primary and secondary data collection, providing justifications for their choices and choosing appropriate techniques to collect quality data for subsequent analysis. Although there was no information about the exact date and time of the data collection, the student included all necessary figures such as pictures, maps, or sample worksheets to explain the methods of the investigation.

Criterion C [5/6]:

The student has successfully collected and presented enough relevant data for the fieldwork question. They have included various maps, graphs, diagrams, and annotated photos to represent the collected spatial data. All figures used by the student include a label and title. However, the cardinal directions and a scale were not necessary for graphs. The student has used a variety of different techniques to present the collected data. Nonetheless, there is no analysis of the sample size, and the maps used are poorly annotated and personalized.

Criterion D [8/8]:

The student has provided a thorough and detailed analysis of the collected data, correctly interpreting it using appropriate descriptive techniques and statistical tests. They have identified significant trends and patterns in the data, as well as important outliers/anomalies and their potential source. The analysis includes references to geographical context and theory, and is directly relevant to the posed fieldwork question and assumed predictions/hypotheses.

Criterion E [2/2]:

The student's conclusion clearly answers the fieldwork question and is supported by the data collected and its analysis. They referred to their original predictions/hypotheses and compared them with the findings. Additionally, they summarized the results of the fieldwork investigation. The student briefly states how the collected data and its statistical analysis support their conclusions.

Criterion F [3/3]:

The student effectively identified the weaknesses of the fieldwork methodology and correctly recognized the potential factors which could have affected the reliability of the collected data. However, there are no strengths mentioned in the evaluation section. The student suggested specific and doable improvements to enhance the fieldwork methodology, which are also well-explained and justified.

Other requirements [0/0]:

The student has successfully adhered to the word limit of 2500 and has numbered the pages in their work. However, it has been noted that the student has not referenced an external source for information regarding the elevation of the river. It is important for the student to include all references to external sources of information in their work.

INTRODUCTION

FIELDWORK QUESTION

To what extent does river Hololo follow the Bradshaw model?

HYPOTHESIS

1. **River discharge increases downstream:** river discharge increase downstream, because of the increasing cross sectional area. In effect, more precipitation - being the main input – increases water volume which affects the river flow.

2. **Occupied channel width increases downstream:** this may be due to lateral erosion in the light of increased volume of water and possible tributaries.

Syllabus relevance

The fieldwork carried out for this endeavor is under option A, Freshwater optional theme. The sub-topic is Freshwater Hydrology and Geomorphology of which discloses how physical processes influence drainage basin system in the light of river discharge and its relationship to stream flow. River processes such as erosion are also enclosed. The theme distils that Bradshaw model suggest how this river characteristics vary from upstream to downstream.

Stream background

Hololo River lies in Butha-Buthe district, it is situated southwest of Liphakoeng with an elevation of 5315 feet, close to Marakabei School. It is a downstream to 'Muela dam, even though because of its sediment deposition problems, the river flow has diverted. In the light of this stream being safe, there are several human activities such as grazing land for cattle, farming and domestic uses; washing. Hololo has a rocky riverbed and displays distinctively three courses of a river –features that make a river safe and suitable for investigating variables such as velocity and depth.

MAP OF HOLOLO RIVER

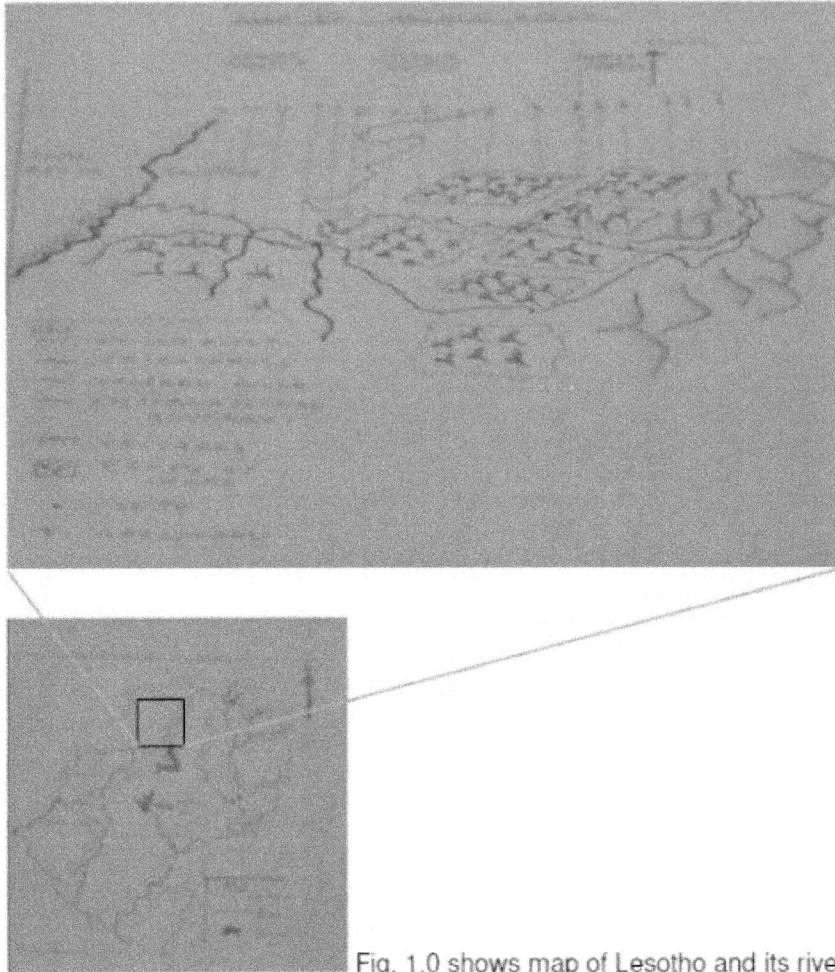

Fig. 1.0 shows map of Lesotho and its rivers.

The Bradshaw model

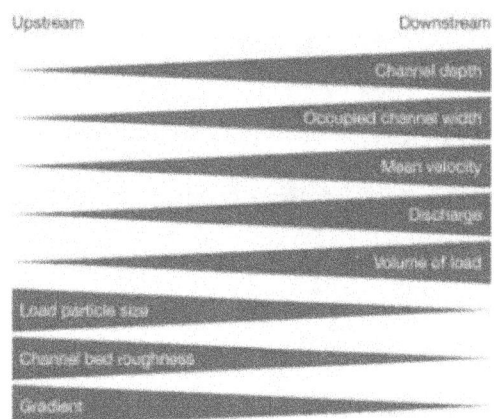

Fig.1.1 shows Bradshaw model exhibiting different relationships of river aspects.

In fig.1.1, the vertex of the blue triangles represent a decrease in a variable, however, its base, which widens, shows an increase in such a variable. In effect, the left side of the model represents upstream, the right representing downstream, thereby, channel width is decreasing towards upstream (broadening of triangle towards left but increases downstream (widening of triangle towards right).

AIM

The motive in this investigation is to prove that flowing from upstream to downstream, there is an increase in occupied channel width and river discharge.

FIELDWORK METHODOLOGY

Systematic sampling method of data collection was used. Starting from the upper course to the lower course, 17 sites were chosen at approximately regular distance between one another and data was collected. This is shown on a map in fig.1.0. This was for the purpose of obtaining equal data for upper course and lower course to allow for comparison of data.

Fig.1.2 shows measuring of occupied channel width in the upper course.

- **River width:** ranging poles were inserted on both sides of the river; up to the occupied channel width. Tape measure is then stretched from one stationary pole to the other across the river to measure width. The width is crucial in exhibiting a trend in occupied channel width and determining the river discharge. This step is repeated 5 times for conciseness.

Measuring stick placed vertically being used to measure river depth; value of depth is taken to be figure on measuring stick aligned with stretched tape.

Load being collected at every interval where depth is measured

Stretched tape measure

Fig.1.3 shows measuring of river depth, channel width and collecting of river load in the lower course.

- **Average depth:** after measuring width, with the tape measure still stretched, distance of width is divided into 10 intervals, at every interval, depth is measured using measuring stick. It is sunk vertically into the river until it reaches river bed, then the height of water is recorded. The depth is essential in determining cross sectional area of a river –which is used in calculating river discharge.

High deposition of matter; pebbles, sand, remains of plants

Float swirling along the river

10m distance measured along the river

Tape measure to mark end point for float to swirl. As float reach this point, stop watch is also stopped for measurement of time

Fig.1.4. shows measuring of river velocity.

- **River velocity:** measure 10m distance along the river in the middle of it. Mark such a distance with ranging poles. Oranges are used as float as they are round and have little weight, this makes them easy to swirl. Release the float gently just before the ranging pole on the upstream site. Measure time taken for the float to swirl from the pole on the upstream site to the other pole on the downstream site of the river.

In effect, the cross sectional area of the stream is obtained by the product of surface width and average depth. Correspondingly, river discharge is obtained by multiplying the cross sectional area with the velocity. This is for the purpose of exhibiting that Hololo stream conform to Bradshaw model of increasing discharge downstream.

The above methodology is repeated in that order along the stream at 17 different stream sites downstream.

QUALITY AND TREATMENT OF INFORMATION COLLECTED

HYPOTHESIS

River discharge increase downstream, because of the increasing cross sectional area.

DATA PROCESSING CALCULATIONS

Sampler calculations for site 5

$$V = \frac{distance(m)}{Average\ time(s)} = \frac{10.0m}{70.6s} = 0.14 \text{ m/s} \qquad\qquad \text{formula (1)}$$

Formula 1: velocity formula is significant as it exhibits how quick a river transports its load at a given time. It also assists in finding river discharge.

Sampler calculations for site 5

Cross sectional area = average depth x width   formula (2)

$$= \frac{(0.70+0.31+0.37+0.26+0.19+0.13+0.16+0.90)}{8} \times 8.83$$

$$= 3.33m^2$$

Thereby:

River discharge = velocity x cross sectional area

$$= 0.14m/s \times 3.33m^2$$

$$= 0.472m^3/s$$

This procedure was repeated for all 17 sites we explored in our fieldwork.

BAR CHART OF RIVER DISCHARGE AND RIVER SITE

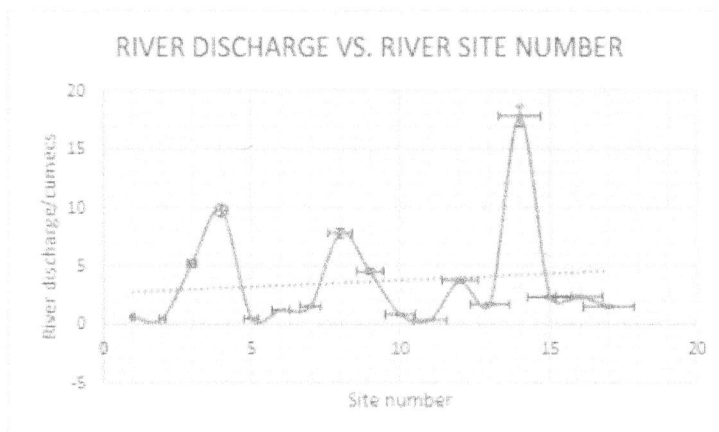

Fig.1.5 shows a graph of river discharge and subsequent river site.

ANALYSIS OF THE GRAPH

The graph shows a trend of river discharge with a corresponding site number. It is convenient to use this graph of nonlinear regression since river discharge increases as site number increase, although some anomalies are encountered. Sites in the upper

155

course such as site 1, 2 and 5 have a generally low discharge. This is due to small cross sectional area of the river due to V-shaped valleys and small surface width of water which inhibit water flow. Because of down cutting caused by rapidly flow of water due to steep slopes in the upper course, these V-shaped valleys are formed. However, there are anomalies in sites 3 and 4 of which there is a high discharge. This is due to high velocity because the oranges freely flowed as the depth of the river may be high at that point. Since it was rainy, this high discharge is also influenced by rain water.

In the middle course, there is an increase in discharge in the light of sites such as 6, 7, 8 and 9. This is due to decrease in friction by the widening of the width and deepening of the depth. Pebbles and cobbles carried by river from the upper course wear down river bed through abrasion, thereby deepening it. Some cracks on river bank allow air and water in, of which due to constant flow of water crack down and widens the river width by hydraulic action. However, there are anomalous sites such as 10 and 11 which is due to friction caused by human activities in the light of a bridge which frequently depreciates thereby some concrete stones falling into the river.

Towards the lower course, there is a significant increase in discharge in the likes of site 12, 13, 14 and 15. This is a result of the river being situated on the flat land without cliffs and large boulders to cause friction. Also, there is a large surface width and river depth that increase the amount of water flowing. Often, in the lower course, there are some small tributaries that increase the river water flow; this may explain the drastic increase in discharge at site 14.

Tributaries are commonly found in the middle course of the river. Not only do they increase river water, but even widen the stream. Similarly, lateral erosion erodes the stream on its sides, thereby still widening the stream. The water surface friction is also high due to less gravity, this in turn widens the river as more water flows on the river sides. Also, meanders, ox-bow lakes, flood plains and levees may also contribute to the increase in occupied channel width.

Occupied channel width is determined by measuring the widening of the river. The measuring tape is used to measure such length across the river up to the point of the river bank where water currently flows.

BAR CHART OF OCCUPIED CHANNEL WIDTH WITH SITE NUMBER

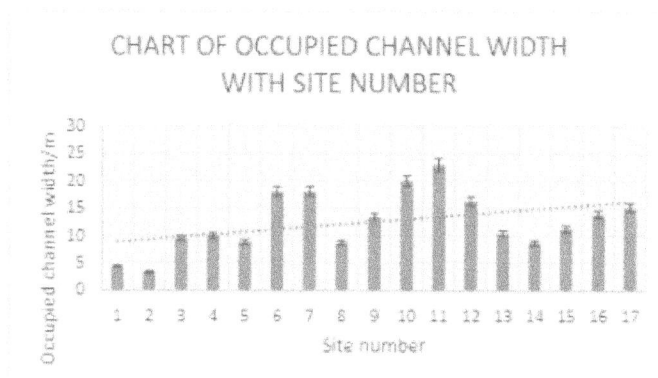

Fig.1.6 shows occupied channel width with site number.

ANALYSIS OF THE BAR CHART

The graph shows an increasing occupied width of the river along with the subsequent site number. As the site number increases; going downstream, the wideness of the river also increase. This is because in the lower course, water near the bank flows at a slow rate due to flat slope, so some water erodes river banks and increase its width.

However, sites 6 and 7 of which are in the middle course and sites 9, 10 and 11 which are in the lower course show an anomalous increase in width. This is due to high lateral erosion that erodes the sides of the river; flooding has widened the river channel by erosion. In the same manner, sites 13, 14 and 15 are anomalous sites as they exhibit decreases in width. That is the impact of high deposition of matter as a result of meander formation. This is annotated in fig.1.4. Less lateral erosion also leads to reduced width. Furthermore, the trend line from the graph is exhibiting an increase in occupied channel width as we move downstream, typical of Bradshaw model.

HYPOTHESIS TEST

For this investigation, t-test is used to test whether the difference between the upper course and the lower course in the light of river discharge and channel width happened by chance, or it is due to stated significant factors. Similarly, the difference is disclosed from two sets of data. T-test compares the mean and standard deviation of the two sets of samples to see if they are the same or different. A value for t is calculated using a statistical formula:

$$t = \frac{\bar{X}_1 - \bar{X}_2}{\sqrt{\frac{s_1^2}{n_1} + \frac{s_2^2}{n_2}}}$$

formula (1)

158

Where x is the value of a measured variable and $\bar{x}$ is the mean value for x, s^2 is square of standard deviation in the data set, n is the number in the sample, and 1 and 2 indicate two different sets of data.

T-test values are then looked up in the t-test table below:

cum. prob	$t_{.50}$	$t_{.75}$	$t_{.80}$	$t_{.85}$	$t_{.90}$	$t_{.95}$	$t_{.975}$	$t_{.99}$	$t_{.995}$	$t_{.999}$	$t_{.9995}$
one-tail	0.50	0.25	0.20	0.15	0.10	0.05	0.025	0.01	0.005	0.001	0.0005
two-tails	1.00	0.50	0.40	0.30	0.20	0.10	0.05	0.02	0.01	0.002	0.001
df											
1	0.000	1.000	1.376	1.963	3.078	6.314	12.71	31.82	63.66	318.31	636.62
2	0.000	0.816	1.061	1.386	1.886	2.920	4.303	6.965	9.925	22.327	31.599
3	0.000	0.765	0.978	1.250	1.638	2.353	3.182	4.541	5.841	10.215	12.924
4	0.000	0.741	0.941	1.190	1.533	2.132	2.776	3.747	4.604	7.173	8.610
5	0.000	0.727	0.920	1.156	1.476	2.015	2.571	3.365	4.032	5.893	6.869
6	0.000	0.718	0.906	1.134	1.440	1.943	2.447	3.143	3.707	5.208	5.959
7	0.000	0.711	0.896	1.119	1.415	1.895	2.365	2.998	3.499	4.785	5.408
8	0.000	0.706	0.889	1.108	1.397	1.860	2.306	2.896	3.355	4.501	5.041
9	0.000	0.703	0.883	1.100	1.383	1.833	2.262	2.821	3.250	4.297	4.781
10	0.000	0.700	0.879	1.093	1.372	1.812	2.228	2.764	3.169	4.144	4.587
11	0.000	0.697	0.876	1.088	1.363	1.796	2.201	2.718	3.106	4.025	4.437
12	0.000	0.695	0.873	1.083	1.356	1.782	2.179	2.681	3.055	3.930	4.318
13	0.000	0.694	0.870	1.079	1.350	1.771	2.160	2.650	3.012	3.852	4.221
14	0.000	0.692	0.868	1.076	1.345	1.761	2.145	2.624	2.977	3.787	4.140
15	0.000	0.691	0.866	1.074	1.341	1.753	2.131	2.602	2.947	3.733	4.073
16	0.000	0.690	0.865	1.071	1.337	1.746	2.120	2.583	2.921	3.686	4.015
17	0.000	0.689	0.863	1.069	1.333	1.740	2.110	2.567	2.898	3.646	3.965
18	0.000	0.688	0.862	1.067	1.330	1.734	2.101	2.552	2.878	3.610	3.922
19	0.000	0.688	0.861	1.066	1.328	1.729	2.093	2.539	2.861	3.579	3.883
20	0.000	0.687	0.860	1.064	1.325	1.725	2.086	2.528	2.845	3.552	3.850
21	0.000	0.686	0.859	1.063	1.323	1.721	2.080	2.518	2.831	3.527	3.819
22	0.000	0.686	0.858	1.061	1.321	1.717	2.074	2.508	2.819	3.505	3.792
23	0.000	0.685	0.858	1.060	1.319	1.714	2.069	2.500	2.807	3.485	3.768
24	0.000	0.685	0.857	1.059	1.318	1.711	2.064	2.492	2.797	3.467	3.745
25	0.000	0.684	0.856	1.058	1.316	1.708	2.060	2.485	2.787	3.450	3.725
26	0.000	0.684	0.856	1.058	1.315	1.706	2.056	2.479	2.779	3.435	3.707
27	0.000	0.684	0.855	1.057	1.314	1.703	2.052	2.473	2.771	3.421	3.690
28	0.000	0.683	0.855	1.056	1.313	1.701	2.048	2.467	2.763	3.408	3.674
29	0.000	0.683	0.854	1.055	1.311	1.699	2.045	2.462	2.756	3.396	3.659
30	0.000	0.683	0.854	1.055	1.310	1.697	2.042	2.457	2.750	3.385	3.646
40	0.000	0.681	0.851	1.050	1.303	1.684	2.021	2.423	2.704	3.307	3.551
60	0.000	0.679	0.848	1.045	1.296	1.671	2.000	2.390	2.660	3.232	3.460
80	0.000	0.678	0.846	1.043	1.292	1.664	1.990	2.374	2.639	3.195	3.416
100	0.000	0.677	0.845	1.042	1.290	1.660	1.984	2.364	2.626	3.174	3.390
1000	0.000	0.675	0.842	1.037	1.282	1.646	1.962	2.330	2.581	3.098	3.300
z	0.000	0.674	0.842	1.036	1.282	1.645	1.960	2.326	2.576	3.090	3.291
	0%	50%	60%	70%	80%	90%	95%	98%	99%	99.8%	99.9%
					Confidence Level						

Fig.1.7 shows t-test values at confidence points. (Anon, 2022)

Similarly, the greater the magnitude of t, the greater the evidence against the null hypothesis. The closer t is to zero, the more likely there isn't a significant difference. There are two important column headings in a table of t-values: 'degrees of freedom' shown as cumulative probability and 'significance level' indicated as confidence points in fig.1.7 above.

Degrees of freedom = $(n_1 + n_2) - 2$, where n_1 is the number of values in sample 1 and n_2 is the number of values in sample 2. After calculating the degree of freedom, that value is matched across the t test values in fig.1.7 to find the confidence level or critical value.

<u>River discharge sampler calculations for t test:</u>

H_0 = river discharge in the upper course and lower course is equal.

H_a = river discharge in the lower course is higher than in the upper course.

$$t = \frac{\bar{X}_1 - \bar{X}_2}{\sqrt{\frac{s_1^2}{n_1} + \frac{s_2^2}{n_2}}}$$ sample 1: upper course,   sample 2: lower course

$n_1 = 8$ $\bar{X}_1 = \dfrac{0.588+0.439+5.199+9.732+0.472+1.201+1.517+7.793}{8} = 3.368$ $s_1^2 = 3.469$

$n_2 = 9$ $\bar{X}_2 = \dfrac{4.509+0.821+0.349+2.764+1.696+17.849+2.319+2.349+1.489}{9} = 4.161$ $s_2^2 = 5.128$

$$t = \frac{3.368-4.161}{\sqrt{\frac{1.469^2}{8} + \frac{5.128^2}{9}}} = -0.376$$

Degrees of freedom = $(n_1 + n_2) - 2$

$$= (8+9) - 2$$

160

Similarly, the greater the magnitude of t, the greater the evidence against the null hypothesis. The closer t is to zero, the more likely there isn't a significant difference. There are two important column headings in a table of t-values: 'degrees of freedom' shown as cumulative probability and 'significance level' indicated as confidence points in fig.1.7 above.

Degrees of freedom = $(n_1 + n_2) - 2$, where n_1 is the number of values in sample 1 and n_2 is the number of values in sample 2. After calculating the degree of freedom, that value is matched across the t test values in fig.1.7 to find the confidence level or critical value.

River discharge sampler calculations for t test:

H_0 = river discharge in the upper course and lower course is equal.

H_a = river discharge in the lower course is higher than in the upper course.

$$t = \frac{\bar{X}_1 - \bar{X}_2}{\sqrt{\frac{s_1^2}{n_1} + \frac{s_2^2}{n_2}}}$$ sample 1: upper course,   sample 2: lower course

$$n_1 = 8 \quad \bar{X}_1 = \frac{0.522 + 0.429 + 5.199 + 9.732 + 0.472 + 1.201 + 1.517 + 7.793}{8} = 3.368 \quad s_1^2 = 3.469$$

$$n_2 = 9 \quad \bar{X}_2 = \frac{4.509 + 0.821 + 0.349 + 3.764 + 1.686 + 17.849 + 2.319 + 2.348 + 1.489}{9} = 4.161 \quad s_2^2 = 5.128$$

$$t = \frac{3.368 - 4.161}{\sqrt{\frac{3.469^2}{8} + \frac{5.128^2}{9}}} = -0.376$$

Degrees of freedom = $(n_1 + n_2) - 2$

$$= (8+9) - 2$$

channel width increases downstream. In the same manner, Bradshaw model also shows

an increasing occupied channel width moving downstream from upstream.

EVALUATION

Weaknesses	Improvements
-Oranges take a while to gain initial velocity, thereby, it took time for water to propel them.	-Using flaw meter to measure velocity may help in getting better results.
-Large boulders in the upper course trapped the oranges, thereby slowing down their velocity and creating anomalies.	-Using another site in the upper course with little or no boulders can limit this error.
-Rainy weather might have tempered with the true river discharge of Hololo because it increased the volume.	-A sunny day is favorable for doing this fieldwork.
-Oranges differed in mass thus, their swirling may have affected the velocity; heavier oranges might have swirled slower than others due to their mass.	-Choosing oranges of approximately equal mass by using mass balance to measure them would decrease such a discrepancy.
Limitations	Improvements
-Tape measure may have not been stretched enough, thereby, the recorded width may be larger than it is.	-Using Pythagoras theorem to measure the river width without crossing it may decrease this error.

-In the upper course, the channel width were too small that the oranges were not freely flowing along the stream.	-Flaw meter is suitable for measuring velocity in wide and narrow parts of the stream
-Crossing the river frequently during variable measuring may have altered velocity measurements.	-Using Pythagoras theorem to measure the width will reduce crossing the river frequently since there would be no need for crossing over the other side.
Errors	
-Random errors may have resulted from the reading of the tape measure and the stopwatch.	-Doing many trials and calculating their mean may limit the occurrence and effect of this type of inconsistency.
-Systematic error might have occurred due to the use of oranges to measure velocity, as the oranges maybe too heavy to swirl freely.	-The use of flaw meter is advisable for limiting such error.
-Parallax error resulted from taking readings of the variables.	-Taking many readings and using modal value might limit parallax error.

BIBLIOGRAPHY + APPENDICES OMITTED

7. EXAMPLE SEVEN (24/25)

Title: To what extent are the streets of Ursynów district of Warsaw sustainable?

Author: M. Lewandowski

Session: May 2021

Level: SL

Examiner's summary

Criterion A [2/3]:

The student has formulated a narrowly focused geographical fieldwork question, clearly stating the location and factors studied during the investigation. The fieldwork question is explored through a collection of primary and secondary data and its subsequent analysis, as well as predictions/hypotheses regarding the results of the fieldwork, which are well-developed and explained. The student has identified and explained how the topic of the conducted study and the areas of the geography syllabus are linked, and included one or more locational maps, which provide clear information about the fieldwork location and include a title, labels, scale, cardinal directions, and a key. However, to improve the analysis, the student should briefly state the precise method of the sustainability analysis and include all the critical information that could help better understand the geographical context.

Criterion B [3/3]:

The student demonstrated a precise and justified selection of methods for both primary and secondary data collection, including the most suitable qualitative and/or quantitative techniques. They also provided clear details such as date, time, and location of data collection, and included all necessary figures such as pictures, maps, or sample worksheets.

Criterion C [6/6]:

The student collected and presented sufficient and relevant data for the fieldwork question, including maps, graphs, and annotated photos. All figures used in the presentation were appropriately labeled and included cardinal directions and a scale. The sample size used for the analysis was detailed and in-depth, with 25 streets from all parts of the district included. The student used a variety of techniques, including personalized and well-annotated maps, to present the collected data.

Criterion D [8/8]:

The student provided a thorough and detailed discussion about the collected data, correctly interpreted it, and recognized significant trends and patterns. They also identified important outliers/anomalies and suggested their potential source. The descriptive and applied statistical techniques are appropriately selected for the data collected and the fieldwork question. Additionally, the student referred to geographical context and theory and the written analysis is relevant to the posed fieldwork question and assumed predictions/hypotheses.

Criterion E [1/2]:

The student referred to their original predictions/hypotheses and compared them with the findings. Additionally, the student summarized the results of the fieldwork investigation, and the conclusion reached by the student is supported by the data collected and its analysis. However, the conclusion did not directly answer the fieldwork question, although the student stated whether the hypotheses are true.

Criterion F [2/3]:

The student effectively identified the weaknesses of the fieldwork methodology and correctly recognized the potential factors which could have affected the reliability of the collected data. However, no strengths were stated. The student suggested specific and doable improvements to enhance the fieldwork methodology, and these improvements were well-explained and justified.

Collapse

Other requirements [0/0]:

The student has met the requirements for the criterion as they have not exceeded the word limit of 2500 words and have numbered the pages. Additionally, they have included references to all external sources of information.

1. Introduction

I have lived in Ursynów, south Warsaw's residential district, for most of my life. Seeing sustainability as an important issue these days, I wanted to assess the sustainability of my place of residence. So, on July 30th, 2020, I went out onto the streets of Ursynów in order to examine the impact they have on the natural environment and the residents.

The definition of the term sustainable street will be based on the following characterization: "Sustainable streetscape ensures that spaces are long-lasting and function as a part of the greater ecosystem employing technologies that manage stormwater runoff and reduce carbon footprint. And it helped create better places for present and future residents".[1]

Based on this definition, I have created an indicator of street sustainability. I remodeled the methodology in *Sustainable streetscape as an effective tool in sustainable urban design*[2] to suit the purpose of my simplified indicator.

Geographic Context

The study was carried out on one of the main streets of Ursynów - Komisji Edukacji Narodowej (also abbreviated and referred to as KEN) and the side streets that connect to it. Ursynów district is located in the southern part of Warsaw, leaning against Kabaty forest, which can in many ways also impact the sustainability of this area. Each side street was measured at different distances from the main street KEN in order to asses its accordance with the sustainability index criteria. Furthermore, the pedestrian flow was measured in select regions of KEN to verify whether higher sustainability yielded better pedestrian experience as was found in the aforementioned article[3].

---

[1] Reeman Mohammed Rehan, "Sustainable Streetscape as an Effective Tool in Sustainable Urban Design," *HBRC Journal* 9, no. 2 (August 2013): 173–86, accessed March 28, 2021, https://doi.org/10.1016/j.hbrcj.2013.03.001.

[2] Reeman Mohammed Rehan, "Sustainable Streetscape as an Effective Tool in Sustainable Urban Design," *HBRC Journal* 9, no. 2 (August 2013): 173–86, accessed March 28, 2021, https://doi.org/10.1016/j.hbrcj.2013.03.001.

[3] Reeman Mohammed Rehan, "Sustainable Streetscape as an Effective Tool in Sustainable Urban Design," *HBRC Journal* 9, no. 2 (August 2013): 173–86, accessed March 28, 2021, https://doi.org/10.1016/j.hbrcj.2013.03.001.

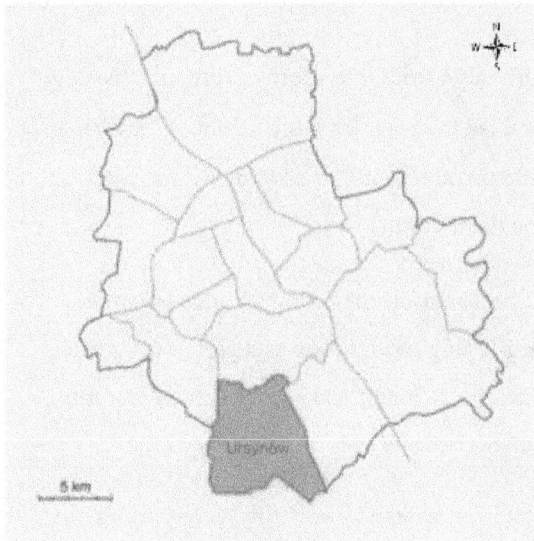

Figure 1 Ursynów on the map of Warsaw
Source: Candidate's own work.

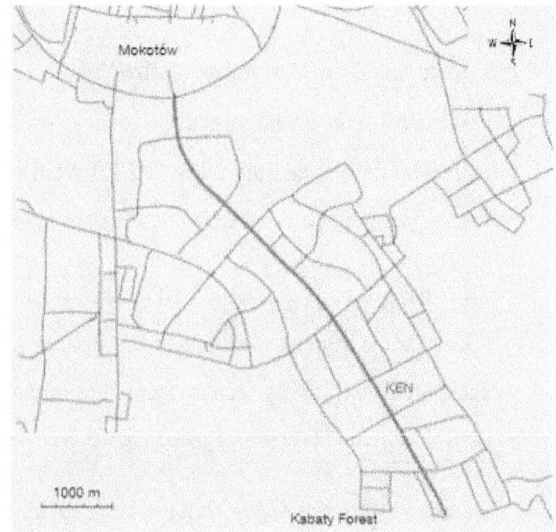

Figure 2 KEN street on the map of Ursynów
Source: Candidate's own work.

Fieldwork question

To what extent are the streets of Ursynów district of Warsaw sustainable?

Hypothesis

The residential side streets are the most sustainable as they offer an increase in the quality of life of inhabitants due to their positive environmental impact.

As the distance from the main street (KEN) increases, the side streets will become more sustainable because of less dense infrastructure.

The side streets with a higher sustainability index ought to attract more pedestrians and thus have a higher result in the pedestrian flow index.

Syllabus links

The fieldwork links to Option G (Urban environments) and specifically to the point about eco city design.

Word count: 356

2. Methods of investigation

2.1 Sampling of side streets of KEN street

KEN spans the whole of Ursynów, more or less creating a straight line and dividing it into halves. Spanning approx. 5.5 kilometers[4], its starting point is located at the northern entrance of Kabaty forest and it ends at the southern edge of the adjacent district - Mokotów district. As one of the main roads in this area, it provides passage for numerous means of transport: Metro Line 1 runs along its entire length, parts of it are covered by bicycle lanes, and some stretches have bus lanes - all of which add to the sustainability of the area.

There are 25 side streets that cross or connect to KEN street. They are as follows: Na Przyzbie, Wąwozowa, Telekiego, Baló, Jeżewskiego, Przy Bażantarni, Belgradzka, Meander, Płaskowickiej, Benedykta Polaka, Indiry Gandhi, Hawajska, Ciszewskiego, Bacewiczówny, Herbsta, Jastrzębowskiego, Zamiany, Stokłosy, Dzwonnicza, Uczonych, Artystów, Pasaż Stokłosy, Bartóka, dojazd do Koncertowej, and Dolina Służewiecka. Below, I will examine and survey these streets with several different indicator values in order to determine their sustainability.

Figure 3 KEN and its side streets
Source: Candidate's own work.

---

[4] "Aleja Komisji Edukacji Narodowej," Warszawikia, accessed August 20, 2020, https://warszawa.wikia.org/wiki/Aleja_Komisji_Edukacji_Narodowej.

2.2 Classification of streets

In order to help with determining the sustainability of different types of streets, I needed to come up with a classification method. After having examined the streets listed above, I came up with the following types of side streets: residential streets, school streets, recreational streets, commercial streets, and thoroughfares. The classification was of my own elaboration based on the *Functional Classification* of *Urban Street Design Guide*[5].

These are defined as follows:

residential street - leading to an area predominately filled by houses or apartment buildings.

Figure 4 Benedykta Polaka
Source: Candidate's own work.

school streets - leading to a school or other educational facility.

Figure 5 Telekiego
Source: Candidate's own work.

---

[5] "Urban Street Design Guide," National Association of City Transportation Officials, accessed December 12, 2020, https://nacto.org/publication/urban-street-design-guide/design-controls/functional-classification/.

170

recreational street - leading to a place predominately designed for leisure activities such as parks, shopping malls, sports arenas, etc.

Figure 6 Jeżewskiego
Source: Candidate's own work.

commercial streets - areas predominately dedicated to office buildings, shops, parking lots, large public transport connection areas (such as metro stations) etc.

Figure 7 Ciszewskiego
Source: Candidate's own work.

thoroughfares - main roads, usually consisting of multiple traffic lanes in one direction.

Figure 8 Przy Bażantami
Source: Candidate's own work.

6

Each of KEN's side streets were assessed at different distances from the main road. The distances established in this study were as follows: 100m, 200m, 300m, 400m, and 500m. Since some of the streets only run through one side of KEN, all the measurements and examinations were conducted on the Eastern parts of the side streets.

An exception to this was Herbsta street which only runs on the western side of KEN, opposite to Jastrzębowskiego street. Thus, it was measured in the same manner as the remaining side streets and was marked on the following map. Hawajska street was much shorter than the measured street length so its western side was also partially accounted for.

Figure 9 Categorization of measured side streets
Source: Candidate's own work.

The table below presents the classification of each side street at varied distances from KEN. The cells were color coded to help with differentiating street types in subsequent parts of the investigation. The color representations are just as on the map above: residential street - blue, school street - yellow, recreational street - green, commercial street - red, thoroughfare - orange.

street	classification at distance 100m	classification at distance 200m	classification at distance 300m	classification at distance 400m	classification at distance 500m
Na Przyzbie	recreational	recreational	residential	residential	residential
Wąwozowa	thoroughfare	commercial	recreational	residential	residential
Telekiego	residential	residential	school	school	residential
Baló	residential	residential	school	residential	residential
Jeżewskiego	recreational	recreational	recreational	residential	residential
Przy Bażantarni	thoroughfare	recreational	recreational	recreational	residential
Belgradzka	thoroughfare	commercial	recreational	residential	residential
Meander	residential	residential	residential	residential	residential
Płaskowickiej	thoroughfare	residential	residential	residential	residential
Benedykta Polaka	recreational	residential	residential	residential	residential
Indiry Gandhi	thoroughfare	recreational	commercial	residential	residential
Hawajska	recreational	recreational	school	school	residential
Ciszewskiego	thoroughfare	recreational	recreational	commercial	commercial
Bacewiczówny	commercial	residential	residential	residential	residential
Herbsta	recreational	recreational	residential	residential	residential
Jastrzębowskiego	recreational	recreational	recreational	residential	residential
Zamiany	residential	residential	residential	residential	residential
Stokłosy	recreational	recreational	residential	residential	residential
Dzwonnicza	recreational	recreational	residential	residential	residential
Uczonych	residential	residential	residential	residential	residential
Artystów	residential	residential	residential	residential	residential
Pasaż Stokłosy	commercial	residential	residential	residential	residential
Bartóka	commercial	residential	residential	residential	residential
dojazd do Koncertowej	commercial	residential	residential	residential	residential
Dolina Służewiecka	thoroughfare	thoroughfare	thoroughfare	thoroughfare	thoroughfare

Table 1 - Classification of KEN side streets
Source: Own elaboration.

2.3 Indicator values

In order to measure the sustainability of the side streets of KEN, a number of indicators were taken into account. The index of street sustainability was created for this purpose. It consists of six categories from which a total number of 15 points can be obtained. The weights of each of the categories remains the same.

criteria	points	explanation
green areas	0-4	0 - no green areas present 1 - green area present on one side of the street, less than 1m width 2 - green area present on one side of the street more than 1m width 3 - green area present on both sides of the street, less than 1m width on at least one side 4 - green area present on both sides of the street, more than 1m width on both sides
bicycle lanes	0-2	0 - no bicycle lanes 1 - bicycle lanes on one side of the street 2 - bicycle lanes on both sides of the street
trees	0-2	0 - no trees growing win the near proximity 1 - trees growing sporadically in the near proximity 2 - area abundant in trees
waste management	0-3	0 - large quantity of litter 1 - medium quantity of litter 2 - low quantity of litter 3 - no litter
road quality	0-2	0 - heavily deteriorated road 1 - lightly deteriorated road 2 - no deterioration of road
pavement quality	0-2	0 - heavily deteriorated pavement 1 - lightly deteriorated pavement 2 - no deterioration of pavement

Table 2 - Index of street sustainability with a total of 15 points
Source: Candidate's own elaboration.

Figure 10 Example of street with sustainability score 14 — Jeżewskiego
Source: Candidate's own work.

Figure 11 Example of street with sustainability score 4 — Na Przyzbie
Source: Candidate's own work.

The primary data for the index of street sustainability was collected during fieldwork on July 30th, 2020. The results were analyzed to calculate the score for each street. The raw scores can be found in the Appendix. Total scores obtained by each side street distance are as follows:

street	total score for distance 100m	total score for distance 200m	total score for distance 300m	total score for distance 400m	total score for distance 500m
Na Przyzbie	4	4	5	4	5
Wąwozowa	9	9	9	11	10
Telekiego	11	14	13	13	12
Baló	13	12	14	12	13
Jeżewskiego	14	13	13	13	14
Przy Bażantarni	14	13	14	13	14
Belgradzka	11	10	11	11	11
Meander	8	9	8	8	8
Płaskowickiej	8	9	11	12	12
Benedykta Polaka	12	12	12	12	12
Indiry Gandhi	12	12	12	13	12
Hawajska	11	9	12	12	12
Ciszewskiego	11	11	12	13	14
Bacewiczówny	10	10	11	13	12
Herbsta	9	9	10	12	11
Jastrzębowskiego	11	12	10	10	11
Zamiany	12	10	11	11	10
Stokłosy	12	12	10	10	11
Dzwonnicza	13	10	11	12	12
Uczonych	10	9	9	12	11
Artystów	12	12	10	11	11
Pasaż Stokłosy	9	11	11	11	11
Bartóka	8	9	10	11	10
dojazd do Koncertowej	12	11	11	11	10
Dolina Służewiecka	11	12	12	12	12

Table 3 - Total scores of index of sustainability for varied distances of each side street.

10

2.4 Pedestrian flow

In order to assess the relation between pedestrian flow and the sustainability of different street types, a random sample of pedestrian flow was taken on 20 different streets and their corresponding distances from KEN. The pedestrian flow was recorded for 5 minutes on each of the chosen streets at 5 pm on July 30th, 2020. In order to get an accurate representation of the flow during the same time period, some of my classmates assisted me in calculating this variable. The street locations were chosen by an Internet random number generator. The locations and their corresponding pedestrian flows were as follows:

street	distance from KEN (m)	pedestrian flow
Na Przyzbie	100	21
Baló	100	18
Jeżewskiego	300	32
Przy Bażantarni	400	36
Płaskowickiej	100	42
Polaka	100	14
Indiry Gandhi	300	28
Hawajska	200	13
Ciszewskiego	100	33
Bacewiczówny	500	23
Herbsta	200	45
Jastrzębowskiego	300	27
Zamiany	500	24
Stokłosy	100	34
Dzwonnicza	200	13
Uczonych	400	17
Artystów	500	15
Pasaż Stokłosy	500	12
Bartóka	300	18
dojazd do Koncertowej	300	25

Table 4 - Pedestrian flow for a random sample.

Word count: 687

176

3. Data processing and analysis

3.1 Assessing the most sustainable street type

The table below presents the average scores and and standard deviations that each street type received.

street type	average index score	standard deviation of score
residential street	10.79	1.84
school street	12.8	0.75
recreational street	10.92	2.50
commercial street	10.78	1.93
thoroughfare	11.36	1.55

Table 5 - Average scores and standard deviations for each street type.

These results were placed on a bar graph.

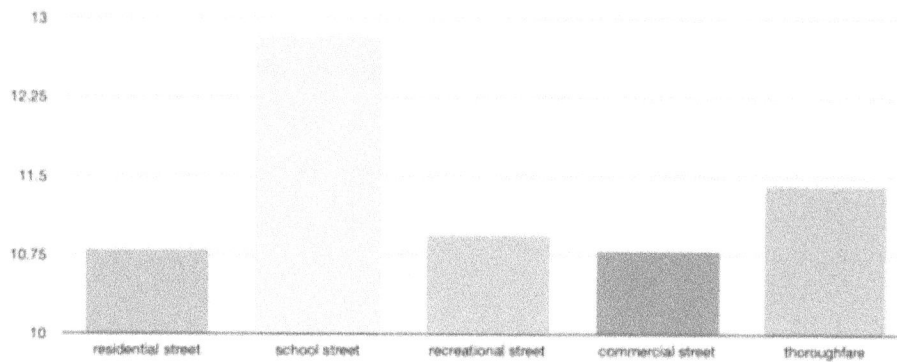

Graph 1 - The average index score for each street type.

It can be inferred from these results that the side street type that is the most sustainable is a school street. It has a significantly higher average score than the rest of the street types. Also, the standard deviation of school streets is low (0.75), which means that there is low variability in these scores and that they are close to each other. The high score of this street type may be due to governmental pressure of keeping streets mainly used by children safe and secure. Furthermore, the low variability of these scores due to the low standard deviation, points to a higher awareness of the maintenance of all streets of this type.

The second most sustainable category of side streets are thoroughfares. Their standard deviation is still relatively low here, although it is higher than in school streets. This shows that thoroughfares, which are generally high traffic areas and routes between certain locations, are also routinely maintained. High quality surroundings of frequently resided public areas may thus be deemed as more important than residential locations.

The remaining street types have similar average index scores and subsequently are the least sustainable. It is quite surprising that residential streets have the third lowest score. This may be due to locations such as schools or high traffic public areas being more prioritized by governmental institutions in terms of their maintenance. However, it is also worth noting that the streets with lower sustainability scores also have higher standard deviations. Even when looking at individual scores of street fragments, one can notice a large discrepancy. Both residential and recreational streets obtained the highest score of 14 and the lowest score of 4. The highest and lowest scores seem to generally coincide with each other in the same street. So, in actuality, recreational and residential streets may not be less sustainable than the aforementioned streets, but certain outliers seem to be lowering the overall score. This may be due to generally lower quality and maintenance of individual streets rather than all streets of that kind.

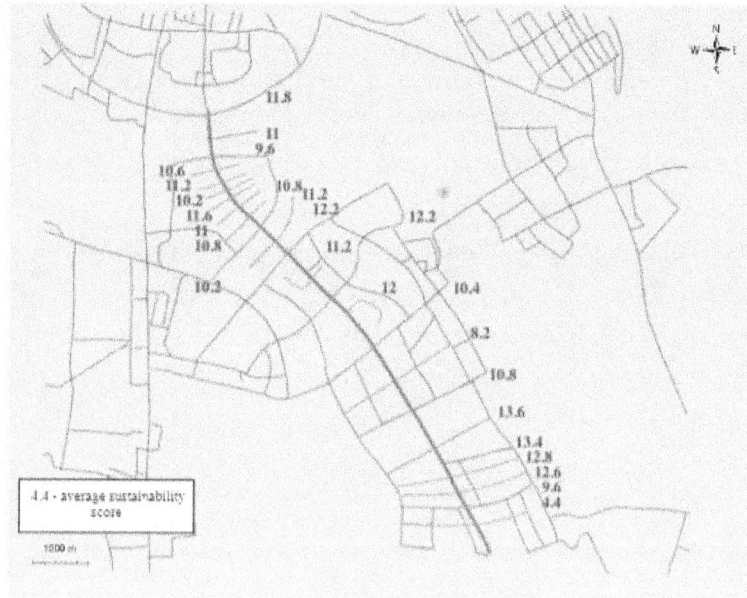

Figure 12 Average sustainability score of each street
Source: Candidate's own work.

Such outlier can be noticed in the case of Na Przyzbie street. The scores of all distances from KEN are substantially lower than those of the following streets. Furthermore, the average score of the whole street in 4.4 which is substantially lower than the remaining streets. Such a result may be surprising as this street is the closest to Kabaty forest. Being close to a large forest, one could naturally assume that this street is more sustainable. However, the street itself is not well cared for. Piles of garbage lie in the surrounding dried up greenery. Fragments of the pavement are broken off in the middle of the street. Such lack of maintenance has thus resulted in an overall lower sustainability score.

Figure 13 Piles of garbage on Na Przyzbie
Source: Candidate's own work.

Figure 14 Lack of greenery on Na Przyzbie
Source: Candidate's own work.

179

On the other hand, the streets with highest average sustainability scores were Jeżewskiego (13.4) and Przy Bażantarni (13.8). A significant factor in the sustainability of both streets is the presence of a park located between them. Przy Bażantarni Park is cushioned between both streets with the highest sustainability score. What is also worth noting is that both of these streets have predominately residential and recreational functions. This therefore points to the idea that both residential and recreational streets may indeed be some of the most sustainable ones.

3.2 The relationship between sustainability and distance from the main street

In order to determine whether the distance from the main street (KEN) influences the sustainability of a side street, average index scores and their corresponding standard deviations were calculated for each distance.

distance from main street	average index score	standard deviation of score
100m	10.68	2.19
200m	10.56	2.00
300m	10.88	1.88
400m	11.32	1.89
500m	11.24	1.86

Table 6 - Average scores and standard deviations for different distances from the main street.

The results were also presented on a scatter graph, to assess relationship between the two variables under consideration: the average index score against the distance from the main street.

180

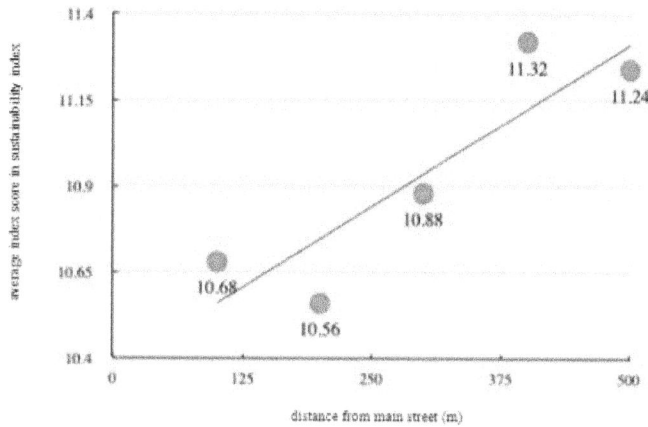

Graph 2 - The average index score against the distance from the main street.

Despite some irregularities on the graph, it can be inferred that the distance from the main street does influence the sustainability of the street. A greater distance from the main street results in a generally higher sustainability score. There are slight discrepancies between the scores, such as the average sustainability index decreasing slightly from 100m to 200m or from 400m to 500m. These distances have small enough differences in the average score for them not to make much of an impact on the entire relation. However, the average score does significantly increase from 200m to 300m to 400m.

3.3 The relationship between sustainability and pedestrian flow

Since the pedestrian flow was measured with a random sample of 20 different streets and their distances from KEN, those streets with their corresponding sustainability index were placed in the table below:

181

street name and distance from KEN	sustainability index score	pedestrians flow
Na Przyzbie, 100m	4	21
Baló, 100m	13	18
Jeżewskiego, 300m	13	32
Przy Bażantarni, 400m	13	36
Płaskowickiej, 100m	8	42
Benedykta Polaka, 100m	12	14
Indiry Gandhi, 300m	12	28
Hawajska, 200m	9	13
Ciszewskiego, 100m	11	33
Bacewiczówny, 500m	12	23
Herbsta, 200m	9	45
Jastrzębowskiego, 300m	10	27
Zamiany, 500m	10	24
Stokłosy, 100m	12	34
Dzwonnicza, 200m	10	13
Uczonych, 400m	12	17
Artystów, 500m	11	15
Pasaż Stokłosy, 500m	11	12
Bartóka, 300m	10	18
dojazd do Koncertowej, 300m	11	25

Table 7 - The sustainability index and pedestrian flow rate for the random sample.

These results were placed on a scatter graph.

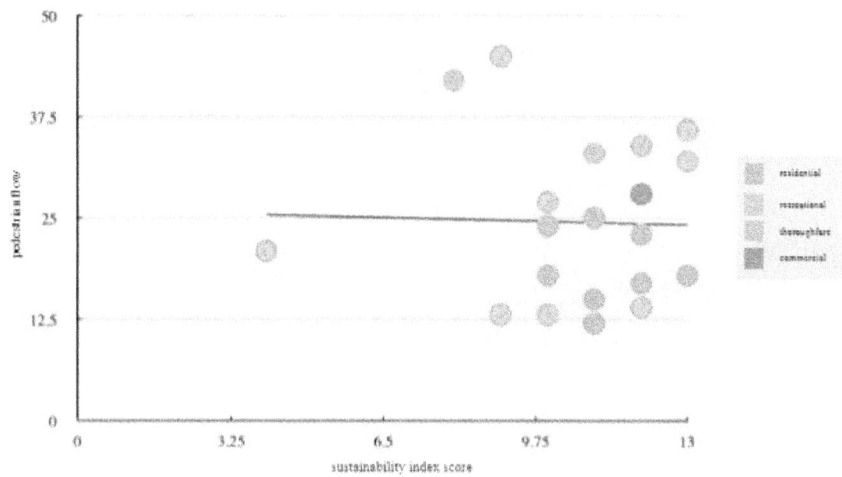

Graph 3 - The sustainability index score against the pedestrians flow for a random sample

These results show that the pedestrian flow is not correlated with a street's sustainability. Streets with a sustainability score between 9 and 13 had pedestrian flow varying from 12 to 45. The lack of correlation could be explained by pedestrian flow resulting more from a street's function than its sustainability.

The majority of streets with the highest levels of pedestrian flow were recreational streets (i.e., Herbsta, Przy Bażantarni, Stokłosy). This could point to higher activity on recreational streets due to their very nature. In particular, Przy Bażantarni and Jeżewskiego have got a large park in their vicinity. Such land use encourages lots of pedestrians to walk in and around the area. Additionally, thoroughfares also had high scores of pedestrian flow which could point to high human activity due to their function as a primary transit route.

Figure 15 Przy Bażantarni park
Source: Candidate's own work.

18

Moreover, the pedestrian flow seemed to increase towards the center of KEN street. This was presented on the map below. Although certain streets are omitted in this rule, in general, the highest rates of pedestrian flow are found towards the center of KEN. The irregularities in this trend may have been influenced by the individual natures of each street (e.g., Polaka street is centrally placed but has pedestrian flow of 14 and it is a residential street).

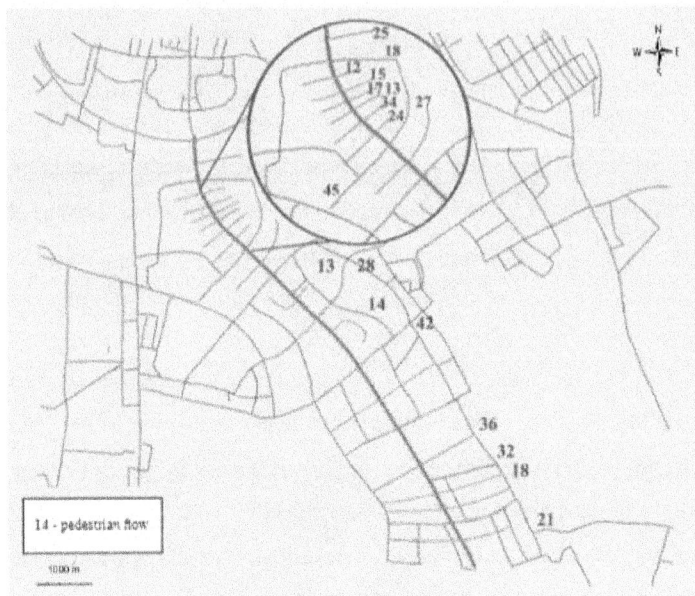

Figure 16 Pedestrian flow random sampling
Source: Candidate's own work.

Word count: 987

4. Conclusion

The first hypothesis stated that residential side streets are the most sustainable. This hypothesis was rejected, following the results of the data analysis. School streets appear to have been by far the most sustainable. This may have been due to the administrative pressure of keeping areas designated for children at high quality.

The second hypothesis stated that as the distance from the main street (KEN) increases, streets become more sustainable because of less dense infrastructure. This hypothesis appears to be true as a positive linear relation is seen in the graph of average sustainability index against the distance from the main street. However, it is worth noting that sustainability for the distance of 200m is slightly lower than that for 100m and 500m was lower than for 400m.

The third hypothesis stated that side streets with a higher sustainability index have a higher result in the pedestrian flow index. This hypothesis was not confirmed, based on the data analysis. There seems to be no relation between the sustainability index score and the pedestrians flow for the random sample. A correlation was found, however, between the pedestrian flow and the street type. The location of the side street along KEN also seemed to influence the pedestrian flow.

Word count: 206

5. Evaluation

It may be hypothesized whether repeating the investigation during a different time period (not heavily influenced by COVID-19) would yield more objective data. Since data such as the pedestrian flow was collected at such time, it may have impacted which side streets were more commonly used. Moreover, the method of selecting the side street fragments for pedestrian flow was imperfect due to the inherent nature of random sampling. A more reliable correlation could be obtained if all fragments were measured at the same time. Of course, an issue that arises is that it would be impossible for a single person to measure pedestrian flow in 125 different street sectors at the same time. In order to objectively measure the pedestrian flow a much larger group of investigators would need to be engaged in the research.

A new, more complex index of sustainability could be created to assess side street sustainability more accurately. The factors that I have considered are those easy to measure without technical equipment. If more advanced tools were available, I could consider aspects such as air pollution levels, soil contamination, and the cleanliness of local water reservoirs. Additionally, more attention could have been paid to aspects such as the appearance of buildings and accessibility of more environmentally friendly options for residents.

The research could be extended by considering more factors such as the relation between pedestrian flow and a sustainability index for each individual street type. Due to limited data concerning certain street types or pedestrian flow rates for each street, such a relation was impossible to determine.

Word count: 262

BIBLIOGRAPHY + APPENDICES OMITTED

www.ingramcontent.com/pod-product-compliance
Lightning Source LLC
Chambersburg PA
CBHW081228020426
42333CB00018B/2463